Praise for

THE SMALL-SCALE DAIRY

"Caldwell doesn't try to convince you that dairy farming is easy (it's not), but she does give you the tools and information you need to make your hard work pay off. From choosing equipment, to caring for your cows, goats, or sheep, to making your own kefir or yogurt, Caldwell covers it all in clear, layperson's terms. An excellent companion for any new or wannabe small-scale dairy farmer."

—ABIGAIL GEHRING, author, *The Homesteading Handbook* and *The Illustrated Encyclopedia of Country Living*

"If you're considering taking your passion for dairying full time, this is absolutely the book for you. Gianaclis Caldwell's *The Small-Scale Dairy* expertly and comprehensively guides you through the entire process of making milk products for home and market. From the science of microbiology and the function of udders, to what's needed to create a milking parlor and even sample floor plans for dairies of all sizes and scales, everything you need to run a dairy is here."

—ASHLEY ENGLISH, author, *Home Dairy with Ashley English: All You Need to Know to Make Cheese, Yogurt, Butter & More* and *Handmade Gatherings: Recipes and Crafts for Seasonal Celebrations and Potluck Parties*

"What a pleasure to read a book about milk production that is at once engaging, nonpolitical, and full of level-headed advice on how to start a small milking operation. I love the practical information on everything from pecking order to water troughs to animal nutrition to finding markets outside the commodity processing system. I can see dog-eared copies of this book becoming staple references in the back pockets of their owners' overalls."

—DAVID E. GUMPERT, author, *Life, Liberty, and the Pursuit of Food Rights*

"Raw milk is today's fastest growing agricultural market, a niche product poised to become mainstream. In *The Small-Scale Dairy*, Gianaclis Caldwell provides those wishing to meet the demand for unprocessed milk everything they need to produce safe, healthy, raw dairy products. From cow and goat care to milk sanitation and testing, Caldwell answers all your questions and provides reliable advice based on years of experience. For both producers and consumers, *The Small-Scale Dairy* is a must read and a valuable contribution to a growing movement."

—SALLY FALLON MORELL, president, The Weston A. Price Foundation; founder, A Campaign for Real Milk

THE SMALL SCALE DAIRY

Also by Gianaclis Caldwell

The Small-Scale Cheese Business:
The Complete Guide to Running a Successful Farmstead Creamery

Mastering Artisan Cheesemaking:
The Ultimate Guide for Home-Scale and Market Producers

THE SMALL SCALE DAIRY

THE COMPLETE GUIDE TO MILK PRODUCTION FOR THE HOME AND MARKET

GIANACLIS CALDWELL

Project Manager: Hillary Gregory
Project Editor: Brianne Goodspeed
Developmental Editor: Brianne Goodspeed
Copy Editor: Eileen Clawson
Proofreader: Ellen Brownstein
Indexer: Lee Lawton
Designer: Melissa Jacobson

Printed in the United States of America.
First printing March, 2014.
10 9 8 7 6 5 4 3 2 1   14 15 16 17

Our Commitment to Green Publishing

Chelsea Green sees publishing as a tool for cultural change and ecological stewardship. We strive to align our book manufacturing practices with our editorial mission and to reduce the impact of our business enterprise in the environment. We print our books and catalogs on chlorine-free recycled paper, using vegetable-based inks whenever possible. This book may cost slightly more because it was printed on paper that contains recycled fiber, and we hope you'll agree that it's worth it. Chelsea Green is a member of the Green Press Initiative (www.greenpressinitiative.org), a nonprofit coalition of publishers, manufacturers, and authors working to protect the world's endangered forests and conserve natural resources. *The Small-Scale Dairy* was printed on FSC®-certified paper supplied by RR Donnelley that contains at least 10% postconsumer recycled fiber.

Library of Congress Cataloging-in-Publication Data

Caldwell, Gianaclis, 1961– author.
The small-scale dairy : the complete guide to milk production for the home and market / Gianaclis Caldwell.
pages cm
Includes bibliographical references and index.
ISBN 978-1-60358-500-2 (pbk.) — ISBN 978-1-60358-501-9 (ebook)
1. Dairy farming—Vocational guidance. 2. Dairying—Vocational guidance. 3. Farms, Small—Management. 4. Dairy farms—Management. 5. Dairy products—Marketing. I. Title.
SF240.7.C35 2014
636.2'142068—dc23

2013048919

Chelsea Green Publishing
85 North Main Street, Suite 120
White River Junction, Vermont 05001
(802) 295-6300
www.chelseagreen.com

CONTENTS

Acknowledgments

How can I not start by first thanking that Neolithic herder, whoever she was, and the first cow or goat that lent her milk for the feeding of early humankind's offspring? Where would we be as a species without this nourishing, sustaining liquid? After that, I would like to thank my supportive husband and daughters for taking over much of the workload here on our dairy, while I once again embarked on a wonderful, educational, but time-consuming literary journey.

Huge thanks to the farmers who inspired and helped: my wonderful Amish friends in Pennsylvania; Art and Teri White of Runnymede Farm, Oregon; Beau Schoch, Schoch Family Farmstead, California; Peter Kindel, Hawthorne Valley Farm, New York; Shari Reyna, Fern's Edge, Oregon; Alethea Swift, Fiore di Capra, Arizona; Suzanne Willow and Lanita Witt, Willow-Witt Ranch, Oregon; Marcia Barinaga, Barinaga Ranch, California; and all others whose milk I sampled, photographed, admired, and was inspired by.

My gratitude also to the critical eyes of and honest comments from my prereaders, several of whom are listed above, but also Pete Kennedy, Esq., president of the Farm-to-Consumer Legal Defense Fund; Tim Wightman, president, Farm-to-Consumer Foundation; Mark Wustenberg, DVM, vice president of quality and operations, Tillamook County Creamery Association; Karin Harris, president of pretty much everything at Highpoint Ranch; Dr. Lisbeth Goddik, Oregon State University; Paul Taylor, Safforésne Farm; and Kirsten Schockey, fermentista, farmer, and soon-to-be-published author. Oh, and of course my primary reader, supportive critic, enabler, and husband, Vern Caldwell.

This book went from a "quick" work (I'll never make the mistake of assuming that again) to a several times over, rewritten book, which I know is better for the effort. This would not have happened if I were not so fortunate to have a stellar editor and superior publisher—with an eye for turning out the best book possible. Thank you, Brianne, Margo, and the committed team at Chelsea Green for continuing to make me look better—and create the best work possible for the sustainably minded, lifelong learning readers.

Introduction

No pastoral scene of country life would be complete without a doe-eyed brown cow being hand-milked by a young maiden—the pail filling with frothy, snowy milk and the cow contentedly chewing her cud. Although this blissful image may be more myth than reality, the milk cow, her small counterpart the dairy goat, and the lifestyle of those sharing their world with these benevolent beasts all have an esteemed place in humankind's agricultural past and present. Indeed, western history has come to identify milk as the embodiment of all that is wholesome and bountiful.

Although the legendary milkmaid may have been replaced by highly automated milking parlors servicing thousands of ubiquitous black-and-white cows and supplying highly processed milk of indistinct origins to the mass market, a new breed of dairy farmer has appeared on the scene. These modern milkmaids and -men might come from varied socioeconomic backgrounds and philosophies, but they have one thing in common: the desire to produce high-quality milk from well-cared-for dairy animals. Whether you are a family in search of self-sufficiency and a source of unprocessed milk; an urban dweller contemplating a couple of miniature dairy goats to supply milk for your cheesemaking hobby; a small commercial farmer tired of selling your milk at regulated prices and barely making a living; or simply a consumer interested in sourcing the most nutritious farm-fresh milk possible, *The Small-Scale Dairy* will tell you everything you need to know about the proper production of nutritious and delicious fresh milk.

My own milkmaid tale began just after I was born in the winter of 1961. My parents had returned to the homestead in Oregon that they had purchased in the late 1940s. After completing chiropractic school in the 1950s and subsequently both opening and closing their practice in Los Angeles, they moved back to their land to farm and begin a family. When I arrived that winter, the structures on the 220 acres included the remains of two homestead cabins, the skeleton of the log house that would become our home, and an outhouse. Conditions were harsh, especially for a woman in her ninth month of pregnancy. It was not surprising that, when I arrived, the usual source of a baby's first milk did not. Fortunately, the valley featured a multitude of small family dairy farms, most milking fewer than a dozen cows. Raw cow's milk from a neighbor's farm became my sustenance.

Our first family cow was a regal Golden Guernsey named Buttercup. My mother says that the day Buttercup arrived, along with her Hereford-cross calf Babe, was the day she felt that the farm was complete. Raw milk, butter, yogurt, and trays of baked custard (which used up the bounty of eggs from our hens as well as the excess milk) were main features at our meals in the log house. When I was thirteen, after much pleading and negotiation with my parents, I got my own cow, a stocky Jersey with quite the attitude. I named her Daffodil. Later I added another Jersey to my 4-H herd, a feminine and sweet girl named Butterscotch. In addition to using it for ourselves, I sold the raw milk, with its thick layer of Yukon gold cream, in gallon jars to many neighbors and friends. I charged $1.00 per gallon, which for a teenager in the 1970s provided what seemed like a good profit.

After becoming a licensed practical nurse, I moved away from the farm in 1984 to begin a new life as a military spouse. My heart broke when I had to sell my cows, horse, and other farm animals. For the next two decades my dairy work was limited to making yogurt from store-purchased milk and rearranging my large collection of cow figurines each time the Marine Corps moved us to a new duty station—having a cow, or even a chicken or two, was out of the question. In 2002, though, we bought a bit of land, and I was ready to "have something to milk." Our younger daughter, Amelia, a wispy eight-year-old, also wanted to have an animal in 4-H, but somehow a six-hundred-pound dairy cow seemed out of the question. Nigerian Dwarf dairy goats, a relatively rare breed at the time, were the answer for our small, two-acre farm.

In 2005 my husband Vern and I, along with our daughters Phoebe and Amelia, loaded up the goats, horses, cats, and dogs and returned to a 25-acre portion of the original homestead to build our small raw goat's milk cheese dairy, Pholia Farm (named after the girls). Within the first year of opening, our product attained critical acclaim from the highest levels of the cheese world, and we have had a waiting list of customers since that time. During the last decade I have learned more about dairying than I ever would have thought possible, including how truly difficult it is to produce superior-quality milk consistently. My travels around the country (and a few trips outside the United States) teaching, speaking, and communing with other small dairy farmers have shored up my own knowledge and perspective—making me feel both confident and reverential regarding raw-milk production.

During the years since we began our cheese company, we frequently have investigated the possibility of selling raw goat's milk, which our state allows. These forays into the particulars of obtaining such a license continued to increase my respect for the process, as well as my knowledge. I visited raw-milk and some pasteurized-milk producers all across the United States, interviewed others on the phone, and attended food conferences and conventions. Even better, I got to taste some amazing examples of wonderfully produced farm-fresh milk from dedicated, hardworking farmers. What an inspiration! Along the way it became obvious to me that a guidebook for the small dairy producer might offer support and guidance to those considering producing milk as well as those already in business. And so *The Small-Scale Dairy* was conceived.

In part I I'll take you on a journey through the history of milk for mankind that will set the stage for understanding where milk production, licensing, and the political scene are today. We'll go over some important things to consider before you decide whether a small dairy is right for you, and we'll cover the many business aspects and considerations that need to be addressed if a small commercial dairy is in your future.

Behind the day-to-day operations of a successful small dairy are the philosophies of the farmers—their motivations, goals, and ideals; the science of dairying—the biology and psychology of dairy animals and the microbiology and chemistry of milk; and the art of dairying—the nuances and practices of milk collection and animal management. Throughout part II, I'll cover the foundational sciences as they pertain to all aspects of dairying, from life at the microscopic level to how animal psychology affects milk production. We'll also talk about the philosophies of animal and land management as they pertain to the production of high-quality milk. The art of harvesting fresh milk combines science and philosophy, as well as practicality.

In part III we'll dig into the nuts and bolts of designing, building, equipping, and maintaining the small dairy. You will also get a chance to apply much of the science learned earlier in part II as we discuss the many different tests that can be done to determine milk quality—and what you can do to improve test results. In the last chapter we'll put it all together for the small dairy business and build a concise, workable risk-reduction plan—an essential element for the small dairy that wants to ensure a long, unsullied career.

Whether you are a home dairy or a small commercial producer, other products besides fresh milk can add value to your diet and income stream. I cover cheesemaking thoroughly in *Mastering Artisan Cheesemaking*, so part IV is dedicated to other fermented products and milk preservation methods. Finally, the book concludes with an extensive appendix, which includes resources, sample charts and forms, and a glossary.

Although not all milk is consumed raw, all milk begins as a raw, unprocessed product. Whether the destination of this milk is a large cooperative processing company, an artisan cheesemaking plant, a group of herdshare owners, or the family table, it can be produced in a conscientious, responsible manner that involves care of the land, the animals, and the product. While economies of scale might limit the large dairy producer from achieving all of these goals, the small producer has the opportunity to take dairy production to a loftier level. Let *The Small-Scale Dairy* be your guide, advisor, conscience, and cheerleader as you create your own modern scene of pastoral bliss and become a modern milkmaid—or man—par excellence.

PART I

THE BIG PICTURE OF THE SMALL DAIRY

· 1 ·

A History of Small Dairying

The small-scale dairy is nothing new. In fact, mankind would not be where we are today without the products that milking animals offered early humans. Meat, hides, bones, horns, labor, and milk made these animals some of the most valuable assets Neolithic humans could own. During the twelve thousand or so years since, livestock products have continued to increase quality of life and shape cultures—from merely enabling people to survive to defining wealth and setting the stage for complex food cultures. Milk itself became a synonym for prosperity and all things good.

Our trip through the history of milk also will take us through a dark time, however, when the creatures who make milk became valued only for the amount of product they could generate, no matter the cost to the beast. We'll see how this and other factors spawned the current public health and regulatory paradigm, one that not only demonizes raw milk, but questions and even ridicules the intelligence and wisdom of anyone who would choose to drink it. I'll introduce you to some of the many manufacturing processes that have evolved along with industrial dairying and how the entire process has changed milk from a simple farm staple to a cheap commodity.

Which Came First, the Cow or the Goat?

Humans did not routinely utilize the milk of four-legged mammals until our primary food sourcing began shifting from hunting for meat and gathering of plants to farming. Interestingly, we may owe our milk-drinking habits to the pursuit of another beloved beverage—beer. Thanks to the analysis of residues on pottery shards from archaeological sites, scientists now believe that the discovery of beer brewing (originally by the unplanned fermentation of wet grains "contaminated" with wild yeasts) inspired Neolithic man to grow even more grain so he could make even more beer. Others believe that grains were first grown to make breadlike products, but either way, both bread and beer either inspired or led to the cultivation of cereal crops about twelve thousand years ago.

As grain production increased and a more stationary lifestyle developed, animals that previously had been hunted in the wild could now be raised in conjunction with the plant crops and harvested for the plethora of valuable products they could provide. Goats and sheep are thought to be among the oldest of the domesticated animals (second only to dogs) and first appeared in Mesopotamia, also called the Fertile Crescent (near modern-day Iran) about eleven thousand years ago. Evidence supports the domestication of cattle about a thousand years after their smaller farmyard cousins, according to Paul Kindstedt in his book *Cheese and Culture* (2012).

While it is easy to think of Neolithic humans as less intelligent than we modern bipeds, their survival depended upon some very basic and highly motivating factors—hunger and comfort. It doesn't take much imagination to see how these ancient people, probably the mothers, would have identified a familiar food source within the udders of their four-legged, furry beasts. Although science is likely never to be able to prove or disprove this theoretical epiphany, I think any parent easily can imagine its likelihood. Nevertheless, at some point milk from these domesticated beasts began to be utilized. Soon the animals became more valuable alive than dead—a basic shift in barnyard economies. A living animal could provide long-term sustenance, with meat and hides becoming additional products from aged and excess young animals. Herds of animals kept for milk, meat, and hides provided status and an early form of wealth assets that animals grown only for meat could not.

Until about fifteen thousand years ago, scientists theorized that almost all human adults were unable to digest milk sugar (lactose). It is likely that the consumption of fresh milk during these early years of agriculture was reserved for needy infants and children, who had not yet grown out of the ability to produce the enzyme necessary to digest lactose. Milk was used in cooking and to produce other products, such as fermented drinks and curdy cheeselike concoctions (perhaps mixed with some of that Stone Age ale). News reports in December 2012 excitedly revealed that tests showed fragments of seven-thousand-year-old pottery sieves were likely used to make cheese. Even older shards of vessels show milk-fat residues, according to Kindstedt, showing that milk was purposefully stored.

Remember that this was an ancient era, a time of no refrigeration and in a region with a generally warm climate. Implements and vessels used to store milk would have harbored some beneficial bacterial, yeast, and mold colonies, creating ideal conditions for the souring and thickening of milk. The naturally clotted milk would have been tangy but refreshing, and perhaps most importantly, our lactose-intolerant ancestors would have experienced no predawn dashes to the privy, as fermentation bacteria would have consumed the majority of the offending milk sugar, turning the previously undigestible milk into a nourishing, better-preserved food that would benefit people of all ages.

Over a relatively short time, natural selection seems to have favored those humans that developed what is called "lactase persistence," the continuation of the production of the enzyme lactase past early childhood, allowing them to drink fresh, unfermented milk as adults. A recent article in *National Geographic* cites the rapid development of lactase persistence as an example of quick genetic adaptation. Without this development, milk as a fresh beverage would likely be a rare product today.

In her fabulous book *Milk: A Local and Global History*, author Deborah Valenze points to a likely theory on why populations farther from the equator are more likely to be lactose tolerant than those from sunnier climes. Although dairying began in the Middle East and Mediterranean, lactase persistence developed first in Northern Europe, likely due to the more northern latitude and therefore shorter daylight hours for a good part of the year. Sunshine is, or was, our primary source of what is now known as vitamin D. Milk fat is also a good source of the same essential nutrient. Without adequate vitamin D a nutritional deficiency called rickets can occur. Under this kind of evolutionary pressure, humans who could ingest vitamin D through milk and milk products and consequently be in better health would have an evolutionary leg up on other, nutrient-deficient individuals.

Although goats and sheep remained the most common dairy animals in the Middle East and the Mediterranean, cows started to gain a foothold in Northern Europe. In regions where grazing opportunities, weather, and evolving traditions favored the larger, less opinionated beasts, bovines became the milk and meat animal of choice. Over time this trend took hold throughout most of the world.

Milk in the Middle

Milk maintained its role as an ingredient in European cooking and food for both young and old throughout most of the Middle Ages. By virtue of their livelihoods, farmers, herders, and peasants were more likely to have access to fresh milk as a part of their daily diets. As the class system advanced, fresh, unprocessed milk developed a reputation as "peasant food." (We'll see in a bit how this image would exert a negative influence on the breastfeeding of infants.) In other parts of the world such as Egypt, India, and Greece, however, milk attained a more revered status, both as a food and as a part of ritual and worship.

During the Italian Renaissance, scholars and elites had enough time on their hands to ponder the innate nature of milk, both in composition and in its effect on the human body. They had some difficulty, however, resolving milk's role in the human diet because the now-discredited Greek theory of "humorism," which had long dictated diet and health, persisted. The "four humors" defined four body systems (blood, phlegm, black bile, and yellow bile) as combinations of hot, cold, wet, or dry. Different foods were associated with each of the humors and were to be consumed to balance bodily functions properly, and thereby, over-all health. Milk, a product of blood, but also cold and wet like phlegm, did not fit easily within this approach and was consequently avoided. Eventually, however, Italian scholars were more driven by their pursuit of novel dishes and cuisine as entertainment and a reflection of social status that they began to take greater interest in milk. In one of the first books on health and living published in 1465 and translated as *On Right Pleasures and Good Health,* milk, cheese, and butter figured heavily in author and scholar Bartolomeo Sacchi's (a.k.a. Platina) advice and recipes. Around 1480 another Italian scholar, Marsilo Facino, in his *Three Books on Life*, even recommended warm milk to help one sleep.

By the seventeenth century most of Northern Europe was regularly consuming milk in one form or another. The popularity of milk as a staple ingredient set the stage for the next commercial jump forward in dairying. In Holland urban growth created an increasing, concentrated population and a consequent demand for milk, cheese, and butter. Natural, fertile meadows and later the drained and fertilized farmland of the nation provided the perfect laboratory for an experiment in what we now think of as industrial dairying. Between the utilization of manure to improve crop yields and the selection of high-producing milk cows (where previously cows often had been dual purpose—providing offspring for both meat and dairy), Holland came to dominate the world of that day in cheese production, fresh milk for its inhabitants, and the export of high-value dairy animals, seen in dairies across the globe today.

It is important to remember that during all of these periods of history, people were practically and intuitively aware that milk was to be drunk immediately; turned into a clotted, soured, or clabbered curds and whey (à la Little Miss Muffet); or made into cheese or butter. Fresh milk poured into wooden and pottery vessels reused for the same purpose would have had many bacteria and yeasts available to quickly assist with the fermentation process, much like the Neolithic pottery containers. Milk was consumed near to its source, and animals had, for the most part, a natural diet and lifestyle suited to their needs. Over time, however, growing urban population centers and consumer demand took milk in an entirely different, dark-ages-of-its-own, direction.

Mobile Milkers: Dairy Animals Abroad

From the time humans started exploring, they often had to take all or part of their food supply with them. When man entered the seagoing era, it was normal for ships to carry animals for meat and milk along with other stores of grains, vegetables, and, of course, ale and its kin. Even the small amount of vitamin C (ascorbic acid) in milk would have been helpful in the prevention of the common sea voyage condition of scurvy. Because goats handled the stress of a sea voyage better than cows or sheep and could survive on a larger variety of forages, they were often the milk-and-meat shipboard animal of choice. It wasn't uncommon for ship captains during the many long sea explorations of the 1700s to deposit breeding pairs of goats on islands—in the hope of returning one day to a well-stocked food source. This practice was sometimes entirely too successful, leading to the eventual deforestation of and loss or diminishing of native species on several islands around the world, such as San Clemente off the California coast and Arapawa off New Zealand.

These Arapawa goats in New Zealand are direct descendants of those first left by sea explorers off the coast of New Zealand. *Image by Geoff Trotter, courtesy of Michael Trotter.*

Dairy cows arrived in North America along with some of the first settlers in the 1600s, and milk supplied both nourishment and comfort during many meals. Settlers quickly learned the value of a cow in providing not only for their own families but as a source of income and barter, not unlike the early animal economies of Neolithic Europe. By the late 1700s there were fairly large-scale dairies (large for their time) of more than one hundred cows near settlements such as Rhode Island. In most cases cows were left to fend for themselves during the winter months and usually faired poorly. But some saw the cow as a future investment and moneymaker worthy of better tending. These more fiscally astute settlers sought ways of increasing milk production and decreasing costs, including through the use of slave labor, by concentrating cows in larger herds, and milking throughout the year.

The growing popularity of milk in the mostly Protestant New World also can be explained in part thanks to its reflection of many Protestant-based values. In *Milk* Valenze cites milk's appeal to many of these settlers through the virtues of its production (hard work) and milk's distinction from the beverages of Europe, namely tea and coffee. In other words, it met with prevailing sentiments in that, as Valenze states, it "nourished without indulgence."

Cows in the Americas by this time had come to symbolize a sort of status symbol that goats did not and so retained their crowns as reigning dairy queens, relegating the goat to a second-class citizen in the eyes of many. The Dutch writer and filmmaker Dr. C. Naaktgeboren challenges the stereotype of the goat as "the poor man's cow" in his outstanding book, *The Mysterious Goat.* Naaktgeboren points out, "What could be more noble than to feed the poor?" Nonetheless, the open spaces of North America (or those that could be opened by deforestation and the removal of Native Americans) continued to invite settlers, their oxen, their family cows, and their milk consumption habits westward.

Urban Disaster—Why Alcohol and Milk Don't Mix

In the 1700s keeping food cool became big business. Thick slabs of winter ice harvested from lakes were cut into blocks and stored in cold houses, to be sold for use throughout the warmer seasons, a practice that became widespread in the eastern colonies and other areas of North America with access to ice. People who had enough money to purchase ice were able to extend their enjoyment of a glass of milk and other fresh products beyond what could be consumed immediately after purchase. Although ice helped keep milk fresh for the well-to-do consumer, the growing urban population included an increasingly large number of people for whom high-quality fresh milk was not a realistic option. At the same time, though, milk played an important role in providing nutrition for the masses, especially infants and children, many of whom had mothers unable to breast feed, often ironically because they were serving as wet nurses to the babies of more affluent women. For these poor folks cow or goat milk mixed with various other ingredients and supplied through a variety of contraptions and methods, was the closest thing to breast milk that could be obtained easily.

As the cities increased in size, dairies either moved farther from the consumers or consolidated into urban dairies. In small townships urban cows could be fed on the lawns of common areas and parks, but as the cities and dairies grew, another feed source was needed. Enter the distillery, the manufacturing plant of another much-desired liquid: alcohol. The distillation of hard alcohol begins with the fermentation of grain. The process creates a great volume of waste grain that, once distilled, is of no more use to the processor. (In chapter 6 we'll talk about conditions under which spent brewer's grains can be safely used as animal feed.) The confined city cows provided the perfect solution. Or so it seemed. In reality the overcrowded, sunlight-deprived cows; the extremely poor-quality feed; the filthy conditions in which milk was collected (including diseases passed unknowingly into the milk by sick and unsanitary workers); and the lack of refrigeration combined for a perfect storm of milkborne illness, disease, and death (not only of humans but cows as well).

By the mid 1800s some estimates placed the population of urban dairy cows in cities such as New York and London at ten to twelve thousand, with most being fed distillery slop and receiving horrendous care and treatment. Even milk produced farther away from the metropolitan areas and transported into the cities was often purposely diluted to increase profits. In the process it was often unknowingly tainted with toxins, chemicals, and disease-causing organisms. Estimates from the late 1880s place infant deaths from bad milk at about ninety-five thousand per year, according to Valenze. Reformers began the cause for the regulation of the industry and protection of the populace.

The Era of the Artificial—Man's Attempts to Outperform Nature

The later part of the nineteenth century saw a large increase in the number of affluent urban families. Prudish Victorian ideals of that time encouraged people to elevate themselves far above the animal kingdom and what were seen as animal-like behaviors—including the nursing of young. As the middle- and upper-class population grew, so did the inability or unwillingness of many mothers to breast-feed their babies. Although the wet nurse (whose own children were most often relegated to what was known as "hand feeding") was the most desirable option for feeding these infants, other class-conscious fears existed, too; namely, that the "essence" of the nurse—her peasant class and lack of "breeding"—might be transferred to the silver-spoon youngster. In addition, humans can transfer their own versions of food-borne illness to an infant. In this case syphilis and tuberculosis were cause for legitimate fears of breast milk. The direct nursing of human babies on animals, especially goats, had been in practice at least since the sixteenth century but was viewed at its best as a medical practice and at its worst as barbaric.

During this same period of time, the field of science was growing in respect, popularity . . . and frankly, arrogance. Trust in scientific studies provided hope for people seeking answers, guidance, and, in the end, better lives. The "scientific approach" also fosters dependence upon what can be proven versus what had previously been known by intuition and tradition. Even in the 1800s, when it was known that babies who were bottle-fed instead of nursed by their own mothers were much more likely to sicken and even die, mothers of all classes continued to seek alternatives to breast-feeding.

Science and nascent market forces studied the properties of human milk and attempted to replicate these same properties in artificial milk for infants—the first baby formulas. In addition to the Victorian sensibilities inspired search for a human-milk replacement, the poor quality of available cow's milk also fueled the quest. The search for a suitable substitute for such a complicated fluid as milk unfortunatly included many infant formula disasters such as vitamin deficiencies in the early 1900s—before vitamins had been identified; vulnerability to illness before antibodies and their presence in the first milk were known; disease causing organisms being passed via the water used to dilute dried baby formulas; aggressive marketing of substandard formula to developing countries; and deaths as recent as this decade that are linked to a rare bacteria that survives in powdered formula and springs to life when rehydrated with water that isn't hot enough to kill it.

For people who are aware of this history, the risks associated with raw milk can seem more acceptable than trust in science and government regulations. While many consumers will regularly purchase the least expensive grocery store milk available, not thinking about the industrial means of production or trusting regulatory oversight to keep them safe, other people feel the opposite. The owners at one farm I interviewed for this book said that many of their customers express a deep-seated mistrust of government and industry, choosing instead to trust to nature and nature's representative: the farmer.

Enter the Entrepreneur—Milk Loses Its Provenance

From its origins as a simple, raw food harvested locally to its current large-scale, multitiered processing and distribution system, milk has come a long way. Milk products such as canned milk, whether plain or sweetened and condensed; dried, powdered milk; and ice cream have played a role in the availability and perception of farm-fresh milk today. With product development comes marketing, with product success comes incentives to increase productivity—or in industry terms, economies of scale.

The transformation from a local food to an industrialized product got a good start in the last half of the 1800s, thanks to entrepreneur Gail Borden. Although milk had been canned in the past, Borden perfected and patented a process for condensing, then preserving the milk in tins. Though bereft of many of the vitamins found in unprocessed milk, canned milk provided nourishment that was portable and long-lasting. Dried, powdered milk, too, became a popular product, both methods proving that convenience and portability were highly desirable, even if that meant that the product barely resembled the original. As Valenze states so well in *Milk*, "As today's marketers well know, a re-created product does not have to taste like its original to be embraced by buyers."

According to Valenze, by the early 1900s the ratio of cows to people had grown to roughly one to five in the United States. Lots of cows meant lots of milk, and lots of milk also meant lots of milk products—and increased competition. The availability of increasingly large processing equipment, refrigeration, and rail transport helped many companies increase production without increasing costs (again, economies of scale). World Wars I and II also played a major role in the development of dairy products produced on an increasingly economical scale and into product lines that could meet a mass, unfocused demand. New substitute dairy products, such as Oleo also threatened to replace a part of the market share previously held by real milk products. To that end dairies often combined resources and formed cooperatives that pooled and processed milk and provided a unified voice for the farmer that sought to both protect and stabilize prices as well as reduce the impact from competing, non-milk products.

With increased competition and lower prices, marketing expanded from developing products to advertising and branding dairy products with pastoral images of the cow and milk from a vanishing era. (The California Milk Board's present slogan "Happy cows come from California" and other such slogans that promote

This 1920s magazine advertisement for canned milk tries to capitalize on the pure image of farm-fresh milk, an approach still used to market commodity milk today.

industrialized milk are nothing new; around 1906 the Carnation Milk Company coined the phrases "the milk of contented cows" grazed on "ever verdant pastures" to draw customers to its red-and-white cans of condensed milk.) But the reality is that by the First World War, in all but the most rural areas, the local dairymaid, her cow, and even more importantly, her fresh, wholesome milk were all but a romantic memory.

With some exceptions, until the mid-twentieth century industry leaders essentially dictated public policy and regulations that governed labor and production. Neither the needless death of babies sickened by often purposely adulterated milk, the grief of their mothers and fathers, the suffering of poorly kept dairy cows, nor the health issues of the dairy workers offered enough impetus for change when up against the profit motive. Reformers campaigned aggressively and persistently for government intervention and regulation. So although today's food regulatory system may seem oppressive to some, it's important to keep in mind that it originated from genuine and desperate need.

The Pasteurization Solution?

Milk of the industrial age was subject to contamination by a variety of deadly disease-causing organisms (pathogens), most of which gained access to the milk through the hands of ill workers, from exposure to poorly sanitized equipment, from diseased cows, and from postcontamination, often from the dilution and adulteration of milk by secondary distributors (to increase their profits). Among the most common milk-borne illnesses were tuberculosis, brucellosis, diphtheria, scarlet fever, typhoid fever, and septic sore throat. Of these, only the first two could originate from the cow herself.

Although disease was a problem for the consumer, the producer lost product and income from spoiling milk and products. Prior to the discovery of pasteurization and the invention (in the late 1800s) of equipment capable of properly pasteurizing milk, as well as refrigeration becoming widely available, other measures were sometimes implemented to prevent losses and keep products from spoiling. To help extend the shelf life of milk, chemicals—that at the time weren't known to be toxic—were sometimes added, including formaldehyde and boric acid. Products containing these toxic substances were intended to help kill bacteria and keep the milk "fresh" and went by harmless-sounding names such as Iceline, Preservaline, and Freeine. At the consumer end, people were encouraged and taught to boil milk, effectively sterilizing it as well as unknowingly destroying many vitamins.

Although microorganisms were first observed using a microscope in the 1670s, it wasn't until the 1860s and '70s that Louis Pasteur and Robert Koch, working separately, correctly described the behavior of microbes, their role in fermentation, their connection to disease, and the fact that they could be reduced or eliminated by temperatures below boiling. Although pasteurization was originally investigated by Pasteur as a way to extend the shelf life of wine and beer, its application to milk and the destruction of milk-borne pathogens soon became evident. The germ theory of disease, which defines microorganisms as the cause of disease, found a receptive audience and fostered a public opinion that all germs were to be avoided and destroyed. The public was subject to this prevailing wisdom through advertisements such as for products like Wanzer's "Germ-No-Milk" and Arnold's steam pasteurizer which promised "the food nearest to nature." This paradigm in which microbes are predominantly bad is still the norm today. In chapter 4 we look at new research, however, that concludes that people actually need a healthy population of beneficial microbes, as well as some exposure to those that can cause illness.

Many people initially resisted pasteurization—the heating of a liquid to a high enough temperature to kill most bacteria but not drastically alter the character of

The advent of the germ theory of disease led producers to promote their milk as free of microbes, thanks to pasteurization.

the liquid (such as boiling and sterilization might)—for a number of reasons. Concerns ranged from objection to the taste and fears that it would lead to more slovenly practices in the milking process to concerns that it would result in the loss of what we now call enzymes and fears about the cost of equipment—and by consequence how that would affect the price of milk for the consumer, and therefore profit.

Other solutions to the milk-related health crisis were attempted, including the establishment of "certified dairies" through which a board of medical experts, called the Medical Milk Commission, provided standards and oversight for the production of superior quality, safe raw milk. Certified dairies became the standard by which safe raw milk was produced on many farms for many decades. Although the Medical Milk Commission no longer exists, the term "certified raw milk" is still bandied about and often meant to mean a dairy producing raw milk under government oversight, i.e., licensed. At the time of its inception, though, the economic realities of bringing the majority of milk produced to consumers already accustomed to inexpensive milk kept most dairies from adopting the certified approach. Pasteurization became the most sensible option for providing safer milk to the mass market.

In 1908 Chicago implemented the first pasteurization laws, followed in 1910 by New York City. In 1914 Massachusetts became the first state to implement government regulation of all dairies and milk sales and was followed two years later by the State of New York. Many states followed the lead of the East, not always making pasteurization mandatory for all milk but often at least regulating and licensing milk producers. In the late 1940s Michigan became the first state to require pasteurization of all milk. Although many states have allowed sales of raw milk within their borders, since the mid-1980s interstate shipping laws have prevented unprocessed milk meant for direct human consumption (as opposed to processing into products) from crossing any borders.

Identity Theft—Milk Loses its Character

Prior to the 1930s the process of pasteurizing milk was accomplished by the holder method (what is referred to now as vat or batch pasteurization), in which milk is heated in a container to a moderate temperature—the current regulatory temperature is 145°F (63°C)—and held for 30 minutes, then cooled rapidly. This method of heat-treating large batches of milk is both time consuming and limited to the size of the vat. It is generally thought to be gentler on the milk, however, and retains more of milk's original flavor and qualities. This method is still common with small-production cheesemakers and a growing number of small fluid-milk producers who supply gently pasteurized, nonhomogenized, full-fat milk.

As population centers continued to grow in the twentieth century, so did the productiveness of the dairy cow—and the economic pressures on farmers. The growing market for fluid milk created a demand that many farmers wanted to take advantage of—creating competition. Economies of scale—increasing size and efficiency to produce a product with less input cost—became the guiding force of the dairy industry. As dairy size and production volume continued to increase, the bottling plant could become more efficient if they could process milk in a quicker, more productive fashion. To the rescue came man's next great milk-machinery innovation: the high-temperature/short-time, or HTST, pasteurizer. In the 1930s this new machinery allowed a continuous flow of milk to be heated to a higher temperature—current regulatory standards are 161°F (72°C)—and held for only 15 seconds, then cooled rapidly. Following the advent of HTST pasteurization, milk could travel from transport truck that delivered from the farm to the milk bottling plant, through the heat treatment, be quickly chilled, then go straight to the bottling equipment in a rapid and efficient fashion. Although this method was too expensive for individual farmers, dairy cooperatives were able to pool their resources and provide bottling and distribution facilities that allowed multiple farms to access this new technology—and get their milk to market within an ever-narrowing profit margin.

By this point equipment that allowed dairy farmers and processors to separate the cream from the milk quickly had been in use for some time, and milk was sold in two simple styles—with cream and without. What was labeled as whole milk varied greatly in the amount of fat it contained—depending on the breed of cows, time of year, and herd management practices. The development of equipment for homogenization—the procedure that changes the size and behavior of milk fat globules so they will no longer float to the top (known as creaming)—opened up new vistas for milk marketing. Homogenization increases the feel of creaminess throughout the milk and saves the consumer the trouble of trying to reincorporate a heavy cream top back into the milk. Homogenization, combined with the ability to test for fat content, allowed dairy industry professionals to remove all the cream, then add it back in at whatever percentage of fat they desired. As a result "whole milk" became a standardized product and excess cream could be diverted toward the manufacture of other value-adding products such as butter and ice cream.

In the 1970s ultra-high-temperature pasteurization, or UHT—heating milk to 280°F (138°C) for two seconds—became an option for milk bottlers. Closer to sterilization, where all life-forms are eliminated (regular pasteurization does not guarantee the death of all microbes), UHT pasteurization greatly extends the shelf life of milk, meaning greater profit margins for the retailer as well as the producer. Organic

WHY THE PRICE OF MILK ISN'T THE COST OF MILK

The best way to begin appreciating the complexity of current milk pricing and the myriad of programs, institutions, policies, and regulations that come together to affect milk prices is to first consider the larger dairy industry. We are talking about the type of milk sold as a fluid or made into products that are meant to be widely distributed and sold in grocery stores across the nation, not locally produced and sold at farmers' markets and specialty grocers. Agricultural products, whether they are tomatoes, corn, or eggs that are a part of this food system, are called "commodities." Before you start thinking of a commodity as a negative term, remember, they exist to supply the nonrural population, and even a large part of the rural population, that does not grow its own food. There will also always be consumers for whom price is more important than quality. For all of these reasons, commodity milk is important! Those of us who are lucky enough to appreciate farm-fresh local foods need to not lose sight of the realities for much of the rest of the population and producers.

Some commodities can be provided easily year-round, such as grains and canned or frozen foods. But milk is considered a "flow commodity"—its production is steady, although its demand varies. As the focus of milk production changed from local to commodity, beginning shortly after the Civil War, dairy farmers suffered financially from this fluctuation of sales. Milk produced on a day followed by a day of low sales, such as a Sunday, might have to be thrown out. Cooperatives, in which many dairy farmers band together and elected leaders to represent them, formed in good part to help negotiate prices that could ensure each farm's survival. Today these cooperatives can cover vast geographical regions, some even coast to coast.

Once the natural variations in sales, and therefore income, were stabilized, other world events such as both world wars and the Great Depression disturbed the price equilibrium once again. Each event inspired a new set of rules, or "improvements," on existing supports—all meant to keep prices somewhat steady for both the consumer and the farmer. What exists now is, as a friend in the industry shared, an almost inexplicable, convoluted, and, in many cases, outdated system of price supports, structures, and controls.

Fortunately smaller, noncommodity producers are mostly outside this regulatory realm. For many farmers this is both a salvation and a new burden—we have to determine our own methods of "price supports," whether that is through value-added products such as cheese or by raising meat animals on the excess; creative marketing such as herdshares or CSA models; or educating the consumer on the high cost of producing noncommodity milk. We can return to the "old way" of living only through new ways of thinking.

milk, cream, and many other fluid-milk products that might not move quickly enough to remain fresh are now commonly treated with this form of pasteurization. In addition, most cream products, such as heavy and regular whipping cream, on the grocery store shelf today are not only UHT pasteurized but also are homogenized and sometimes include ingredients such as thickeners and sweeteners.

It was not only mechanical inventions that encouraged and allowed for uniformity of commercially produced milk. Practices such as the pooling of milk from multiple herds of increasing size and the standardization of herd management continue to play a role today in producing predictable milk. Dairy science courses help farmers learn to manage and feed their herds in a standardized and efficient way. Feeds are carefully analyzed so farmers can balance their cows' diets with what are believed to be the optimum nutrients for high milk production and health. Grazing on forages that vary greatly by season and location has become a rare practice, for reasons that include undesirable variations in flavor and the economic realities of maintaining enough pasture and moving large numbers of cows efficiently from pasture to parlor.

My husband, Vern, still calls whole store-bought cow's milk "vitamin D milk." When he was a kid (not growing up on a home dairy as I did), whole milk fortified with vitamin D began appearing on the grocery store shelves. His folks had been victims of the "fear of fat, especially milk fat" movement of the '60s and '70s and typically mixed the family's "milk" from boxes of powdered nonfat dry milk. So as a treat he would save up his own money to buy a gallon of "vitamin D" milk for himself.

Beyond vitamin D, which a period advertisement I found described as "added sunshine," other nutrients have been added to fluid-milk products, including vitamin A to skim or partly skimmed milk (to replace the natural vitamin A removed along with the milk fat); calcium; omega-3 (a short-chain fatty acid)—produced by feeding the cows fish meal or flaxseed; acidophilus, a probiotic bacteria to assist with gastrointestinal health; added fiber; and the enzyme lactase, which makes milk digestible for the lactose intolerant. With all of the research and effort that has gone into improving milk for the supposed sake of our health, I wonder why one of the arguments against raw milk is that people should not be concerned about the nutrients lost during processing, as those are easily obtained from other food sources—a statement I actually think is valid, but one that suggests that these efforts to make and market mass-produced milk as a major source of nutrition is dubious.

Beyond things added and removed, much research has gone into some even more bizarre means and methods to produce milk. Fairly recently the engineered growth hormone rBST, marketed in the United States as Prosilac, was being widely used to push highly productive dairy cows to new milk volume horizons. Animal health issues, including mastitis, hoof problems, and reproductive issues led many countries to ban the use of rBST. Despite being approved by the FDA, the usage of rBST in the United States has fallen against popular outcry, with many dairy cooperatives and product manufacturers labeling their milk rBST-free. The latest bizarre direction for our old friend milk comes in the form of the FDA's approval of the use of milk and meat from cloned animals. At this time cloning is extremely rare and not very successful, but still, it makes you wonder what is ahead for both our dairy animals and us.

Most commercially produced milk today is the offspring of centuries of market pressure; fear of microbes and a "safety first" mentality; and nutrient manipulation to compensate for processing—and to tap into consumer fads. Small-scale milk producers and their eager customers represent a body of people with a growing desire to retake food rights, consume minimally processed products, and remove their meals from the long chain of handling and manipulation. With knowledge and information we can retake milk and return it to the simple, unprocessed sustenance that our Neolithic ancestor so astutely identified as a valuable food source.

·2·

Is the Small Dairy Right for You?

Locally produced foods, especially those available directly from the farmer, are increasing in demand all over the country. Motivations such as reducing the distance food travels from its source; supporting local economies; reducing the number of intermediaries between the farm and the table; concerns over multistate food-borne illness outbreaks from mass-produced and -distributed products; and a desire for maximum flavor and nutrition all play a role in the growing interest in local food supplies. Small-farm-produced dairy products are equal in popularity to veggies—but are much harder to find. Whether it is raw milk bottled by a farmer milking a few cows; artisan cheese made from single-source herds raised on local pastures; or high-end gourmet ice cream, yogurt, butter, and kefir, these types of dairy products offer both farmers and consumers an improved quality of life. For the consumer who also wants to be his or her own producer, a single cow or a pair of goats or sheep can turn a hobby farm into an artisan dairy. But the life and realities of any dairy producer are not for everyone.

In this chapter I will cover some of the great reasons to produce and sell milk—either directly to the consumer or to an artisan producer—as well as the realities. I'll give you some examples of small dairies, from the city farm with a pair of miniature goats in the backyard to the small commercial dairy milking a few cows. We'll talk about how your choices might be limited by zoning laws and other important factors such as sourcing affordable feed, handling animals, and processing the different waste products related to milk production. Finally, I'll delve into how to choose your species of milk animal, including a brief overview of suitable breeds for a variety of situations and tastes.

A jar of farm-fresh raw milk promises a taste and nutrition experience that appeals to many people.

Why a Small Dairy, Why Now?

As we learned in the last chapter, milk is a highly valued food staple. By producing minimally processed, lovingly harvested milk for your family or your community, you link your animals and land to your life and to the lives of others in a fundamental way. When I was a teenager hoping to have an animal project as a part of the 4-H program, I wanted that animal to feed me—not through its death, but through its life. Dairy animals provide long-term sustenance and a connection to the life cycle that brings the dairy farmer more than just milk. When that milk is shared with others, they too can connect to that link. Today that connection is more relevant than perhaps ever before—not out of necessity, but out of desire. Today's dairy farmer may provide for just his or her family; to multiple families through a herdshare; at the retail level with raw or even gently pasteurized, nonhomogenized milk; or by selling his or her milk to an artisan cheesemaker.

You would almost have to have been living under the proverbial rock to not have heard about the popularity of raw milk—and the controversy. As a long-time lover of raw milk and of science, I find it impossible, and personally unnecessary, to boil the debate down to a black-and-white, clear-cut right or wrong. Although many states do not allow the sale of raw milk as a final product, no state disallows the consumption of raw milk from your own animals. And all but one state, Maryland, follows the FDA ruling regarding the production and legal sale of properly aged cheeses made from raw milk. And they are currently allowing some production to occur. In many states that heavily restrict the sale of raw milk, pressure from organizations and individuals is helping make food rights a progressive issue. In the resource section of this book, I include contact information for groups providing support and information regarding raw-milk rights.

Where raw milk is legal to sell, whether that be directly from the farm, through a herdshare, or at the retail level, prices average three to five times that of conventionally produced milk. The input costs of building a facility to produce milk vary tremendously based on what level the sales will occur at (from direct on-farm only up to the retail level) and the restrictions and requirements of the state. Interestingly, the price the consumer pays seems not to reflect the input and ongoing costs of the farmer, but rather what the market will bear—with consumers located near large metropolitan areas and regions with a more progressive population willing and able to pay the most; for example, some of the highest raw-milk prices I encountered were in and around the California San Francisco Bay area.

In areas where only pasteurized milk is legal for sale, whether that be on-farm or full retail, or where a demand exists, farmers can still increase their profit margin and provide a superior product, when compared to that available in most grocery stores, through the sale of nonhomogenized (also called cream top), low-temperature-pasteurized milk. Milk processed in this low-technology small-batch fashion can command close to the same price as raw milk. Input costs will be higher, though, because of the steep cost of even the smallest pasteurizer. In chapter 9 I'll talk about equipment options for pasteurizers and small bottling equipment—often required by some states for retail milk sales.

The artisan cheese movement has created a demand for small volumes of artisan-produced milk. Cheesemakers and crafters of other dairy products who don't want to milk their own animals often have a hard time finding a superior source for small volumes of raw milk. These artisans are looking for milk that reflects *terroir*, the uniqueness of the species, and the farmer's attentive care of the animals. Although cheesemakers rarely can pay close to what the retail customer expects to pay, they might be the perfect customer for consistent sales of a bulk volume of milk—without the need for bottling, labeling, or processing. So although the price may seem low, there is far less labor and input required on your part. I frequently advise prospective cheesemaking operations that locating a milk source may be their most daunting hurdle. If you decide to sell milk to a cheesemaker, plan on entering into a contract (there is a sample in appendix C) that will lay out details such as price, bacteria counts, somatic cell counts (don't worry, we'll cover all of these later in the book), and butterfat levels.

How Small Is a Small Dairy?

The small dairy can be any size that still allows you to feel personally engaged in the animals and the process. For some people that may mean only one or two cows or goats; for others that might mean fifty or more. When we have visitors here at our farm and take them out to hike with the goats while they browse, people are surprised that we can recognize nearly a hundred goats—and know their names. I remind them that they likely know far more than a hundred people whom they can recognize on sight. To the farmer the animals are not only "people" we know; they are our business partners.

The smallest dairies are found in the backyards of suburban and even urban farmers. Fortunately, it is not uncommon to find ordinances that allow city farmers to grow not just fresh tomatoes and carrots but also to produce eggs and milk. For milk, miniature breeds of goats are the only choice but easily can produce enough for a small family. Probably the biggest hurdle facing the urban dairy is finding a suitable, available father for the goat's young—as full male goats, like their chicken counterparts, roosters, are never allowed. For more tips on setting up an urban goat farm, I recommend the book *City Goats* by Jennie P. Grant.

For those living on one or a few acres, and where zoning laws allow residents to keep livestock, a single cow, a couple of full-size goats or sheep, or a handful of small-stature goats might constitute a small dairy—either for home and hobby or for a tiny commercial enterprise. Small acreage brings with it challenges, but none that cannot be addressed through proper planning and creative strategizing.

If you want to invest in a small commercial facility that will feed many families, however, you'll need to do more planning to make sure that the business is a viable

SMALL DAIRY PROFILE: NOVELLA CARPENTER, OAKLAND, CALIFORNIA. URBAN HOME DAIRY.

A few years ago I visited a tiny dairy in the most unlikely of places: Oakland, California. And not a gentrified, upscale neighborhood, either, but rather the tiny agrarian kingdom carved out of a rugged, illicit-activity neighborhood by author Novella Carpenter and her partner Bill Jacobs. Novella's *New York Times* best-selling book, *Farm City*, is a gritty yet poetic memoir of her journey to urban farming and the animals she loved and lost—to the dinner table—along the way. Her dairy goats, two adorable Nigerian Dwarf goats named Bebe and Ginger, happily utilized the stairway from Novella's upstairs apartment down to the postage-stamp-size yard as a part of their habitat. When I sat down with Novella that day, we sampled several of the delicious cheeses she had made from the milk of her city goats and talked about the couple's plans for buying the vacant lot next door that was serving as her garden and orchard.

While Novella's farm story is unique, urban dairy goats are becoming more common as cities change ordinances to allow for the keeping of small livestock. Novella made use of her small space by seeing to the goats' needs through utilizing features such as a stairway to increase the animals' access to activity and their need to climb. Her front porch doubled as hay storage, her back porch as a milking parlor, and her kitchen as the milk house. Vegetables and trimmings from her garden helped reduce the feed bill, and manure from the goats helped nourish the garden. I had no doubt that her model was possibly more sustainable than many farms a hundred times the size (which would still be fairly small). To learn more about Novella's city farm, visit her blog at www.ghosttownfarm.wordpress.com

Novella Carpenter with her two Nigerian Dwarf dairy goats, Bebe and Ginger, at her urban farmstead in Oakland, California.

prospect. A dairy sized to create a decent income flow might be as small as a few cows to a couple of dozen or a handful of goats to several dozen—depending on the expenses that the farm income must cover. As you go through this chapter, there will be opportunities to analyze and collect data that will help you create a business plan—and make decisions regarding your dairy. There is no one-size-fits-all plan or income and expense estimates, so you must be prepared to do some homework. If you already have a dairy farm and are investigating the options for selling milk directly to customers, then it will be easy for you to address many of the topics in the rest of this chapter and develop your plans.

Personal Suitability

In my opinion, assessing your personal suitability for a hobby or career in small dairying is one of the most important preliminary tasks—and also one of the least explored. In an article she wrote for *Wise Traditions* (the periodical published by the Weston A. Price Foundation), Oregon herdshare dairy woman and small dairy mentor Charlotte Smith shared an example of the kind of reality check that can occur after opening a small dairy.

"I came up with the most brilliant idea," wrote Smith. "I'd get a cow, milk her, and sell the surplus." She shared her family's raw-milk budget, then said she found it simpler and possibly more affordable "to pay three hundred dollars per month for your raw milk than get into the raw milk micro-dairy business." Fortunately for Smith, her personality suited the job, and she emphasized that it has been fulfilling and overall worth the effort. Before you jump into the microdairy pool, let's test the waters by taking a serious look at financial and lifestyle suitability as well as the emotional and physical realities of running a small dairy.

Having a small income-producing farm of any kind is what I refer to as a "lifestyle business" rather than a growth or investment business. A dairy is also a commitment to a level of work that has no time clock—you are always working or on call and with arguably more demanding year-round obligations than you would have with other livestock enterprises. It is also not a type of work that is widely lauded for the production of wealth—at least not the kind that you can save in the bank. You must gain a sense of personal satisfaction from the work that will sustain you through many stressful times, not the least of which will likely relate to finances. When dairy farmers produce, package, and sell milk from the farm, rather than trucking it to a processing facility, they add several more layers of responsibility to an already robust workload. To be a moderately to highly successful dairy farmer (and by that I mean producing an income that supports the farm lifestyle), you need to have a comfort level with the business aspects of the work, including balancing accounts, paying bills, dealing with invoices, and planning for big annual or semiannual bills such as feed and insurance. Even if you don't enjoy this type of work, you need to be able to perform it. Most farmers would much rather be outside fixing fences, monitoring animal and pasture health, or baling hay than sitting at a computer e-mailing reminders to those who haven't paid their bills.

The emotional reality of dealing with animal lives and deaths is another potentially overwhelming factor that is impossible to describe or prepare for adequately. If you have a very small farm with only a few animals, your encounters with these harsh realities may be few, but as your herd size increases, so do the odds of dealing with illness, tragedy, and tough decisions. One spring we lost 20 percent of our newborn baby goats to an intestinal illness caused by a protozoa called coccida. Their deaths were not quiet or peaceful, and at that point there was nothing we could do to save them, only lessen their suffering by euthanasia. I spent the entire year on edge, fearing the next misfortune. Successfully managing the ongoing challenges of a farm includes many tough lessons and unavoidable mishaps that you must learn to take in stride if your farm is to survive for the long haul.

The physical challenges of working with livestock and running a farm are commensurate with the farmer's physical condition. In my consultation work, often the most difficult advice I have to give people involves their age and physical limitations. Many people are drawn to this work and are financially ready to enter into it when they are retiring from another, perhaps more lucrative, career but are also at an age when physical limitations become more apparent. In this case, if you have enough money and accept the need to hire help, then the dairy is still a possibility, although you still need to consider whether it will be physically and financially sustainable in the long run. The physical work of a dairy occurs in seasonal bursts that can make it even more difficult, from dealing with new mothers in milk and their rambunctious youngsters in the spring to bucking, hauling, and stacking hay in the hottest months. I encourage you to chart your future years in the business with an eye toward hiring help and paying for more labor as your age increases.

Loading and unloading hay is tough, seasonal work—in the hottest part of the summer.

Input and Output—Feed and Waste Realities

Once you have determined that you and the small-dairy lifestyle are compatible, I suggest taking a look at the realities of providing feed and dealing with the waste produced by even the smallest dairy. Before doing any market research or designing a barn, you need to find out if feed is available at a price and volume that works and if your land and plan can deal with the manure, bedding, feed waste, wash water, and land damage that is part of having a dairy.

THEN YOU MIGHT BE A DAIRY FARMER

1. Does your idea of a vacation involve getting up several hours earlier than normal to finish the chores in time to attend a raw-milk educational conference, then arrive home late, do chores again, and still get up on time the next morning?
2. Can you give hours of care and love trying to save the life of a weak bull calf and still come to terms with his being served as steak in a year or two?
3. Do you find that the rank, acrid smell of male goats in rut makes you think fondly of cute baby goats in the spring?
4. Does getting ready to go to town include changing into clean Carhartts?
5. Do you routinely pay more for animal feed than for your own dinner out?
6. Do your friends often point out that you have a bit of alfalfa stuck in your hair?
7. Do you find inserting your arm into a laboring doe's uterus to untangle triplet goat kids an interesting challenge?
8. Can you make the decision to euthanize a suffering animal, perform the job, feel sad, and still sleep well that night?
9. Does your idea of a balanced workout include doing squats while working in the milking parlor?
10. Does producing wholesome food and feeding your community make you happy?

It took us a couple of years to secure a regular, reliable, and superior source of hay for our goat dairy. Even though we live in a great region for growing hay, most of the larger producers already had buyers lined up for their crops. We had to source it from several different growers until we finally got our foot in the door and connected to a steady supplier. The larger your dairy, the more likely you can find a source, as it is far easier for hay growers to sell most of their crop to one buyer for a single payment. If you are a very small farm, you will likely not buy an entire year's supply of feed at one time, giving you an advantage in staggering the costs over the year but the disadvantage of paying a premium price and likely varying degrees of quality.

If your farm includes land for year-round grazing or the possibility of growing your own hay, you may find yourself at great advantage in regard to feed supply, but you also will have other costs and issues. You'll need to invest in equipment, maintenance, and labor to maximize the land and its potential to feed your stock. From tractors (or horse or oxen power) to electricity for irrigation and the labor of moving irrigation pipes, managing crop-producing land can be a job unto itself.

In addition to the cost of fodder (hay and pasture), you should factor in other feed costs for supplements—minerals, vitamins, and so on—and concentrates (grain). The expenses for these parts of the dairy animal's diet can be surprisingly high, especially when your need goes from feeding a couple of goats a cup or two of grain a day to feeding a herd a 50-pound bag of expensive, organic oats a day. You also should source and price bedding, such as straw or wood shavings, for the animals. And don't forget to factor in cost estimates for proper storage facilities for feed and supplies!

Once you have verified a sustainable feed source for the herd, it is time to face dealing with what remains of the feed after the animal has processed it—manure and urine as well as soiled bedding and wasted feed. If you are a very small

Dairy water buffalo are a rare sight in the United States. These cows provide rich milk at Ramini Mozzarella farmstead in California.

farm, you won't have much trouble utilizing these wastes as fertilizer, compost, or mulch. You may even find neighbors and customers for some of these valuable plant nutrient sources. Keep in mind the innate differences between very wet cow manure and the dry, pelletlike feces of goats and sheep. These differences each present possibilities and obstacles when used as plant nutrients.

Depending on where you live, you may encounter many regulations regarding animal waste—and often for good reason, since the same nutrients that can enhance soil also can contaminate groundwater and waterways when not applied properly. Although these regulations and the requirements they put upon the producer can be a burden, they are meant to help protect the land and keep people healthy. If you are planning on building a commercial facility, even a small one, be sure to check local zoning laws and with your state department of agriculture (or its equivalent) for relevant regulations. In many states, including ours, even the smallest facility is likely to be required to obtain a Concentrated Animal Feeding Operation (CAFO) permit. You probably don't have what this permit was named after—a feedlot— in fact I believe these permits might better be called animal waste-management permits. They require knowledge of waste nutrient content and a plan for how this will be managed on your land. Obtaining them is an onerous process, but one that is also educational and might serve your land in the long run. See appendix A for more information.

In addition to waste from the animal side of the dairy, your dairy processing areas will create wastewater containing varying levels of cleaners, sanitizers, milk residue, and perhaps a bit of manure from the milking parlor. At the home scale this can be managed easily by your existing sewage or septic system—especially by choosing eco-friendly cleaning and sanitizing chemicals. At the larger scale, or at any commercial dairy, this wastewater will likely need to be managed by an inspected and approved system. It might be a traditional septic or greywater management system designed to hold the water until the chemicals stabilize to the point that it can be safely spread on crops.

Dairy sheep are still relatively uncommon in the United States, but they produce milk that is highly sought after by artisan cheesemakers.

Cow, Doe, or Ewe: Which Is Right for You?

If you have passed the previous suitability tests, then congratulations! Now you can spend some time pondering a more enjoyable topic: What kind of milk animal would best suit your needs? For barnyard matchmaking to be a success, you need to address several concerns. The volume and components; innate flavor and aesthetics of the milk; your land, including available feed, topography, weather, size, facilities, and neighbors; your personal disposition; market issues regarding not just the milk but also surplus animals; and the availability of good starter stock.

Volume and Components

A rule of thumb can be applied regarding the richness of milk; in other words, how much fat it contains. The smaller the animal, the less diluted the milk and therefore the higher the fat. In dairy cows, goats, and sheep, the largest breeds also have the lowest milk fat, as a whole, while the smallest breeds have the richest. Of course, there always will be exceptions to this rule, but it can be applied as a reliable standard. Even within breeds, typically the smaller animals will have more condensed milk (naturally, without the need for industrial milk-condensing equipment). With more concentrated milk—higher in both protein and fat—cheese yields will increase, as cheesemaking might be simply thought of as the removal of water from milk. If you prefer a naturally less concentrated milk, then choosing a larger breed is a good idea, although you can also simply dilute the milk to your taste!

A high-quality dairy goat in her prime (one of the big girls, not one of my Nigerian Dwarfs) may give up to 2 gallons of milk each day for the first several months after she gives birth. A hard-working Jersey cow at her peak will give at least three times that much milk, while a superior, tall for her breed Nigerian Dwarf goat will max out at about ¾ gallon. Typically, cow and goat lactations (the number of days that they are milked from the time of birthing until the dry period before the next baby is born) are measured over a ten-month period (305

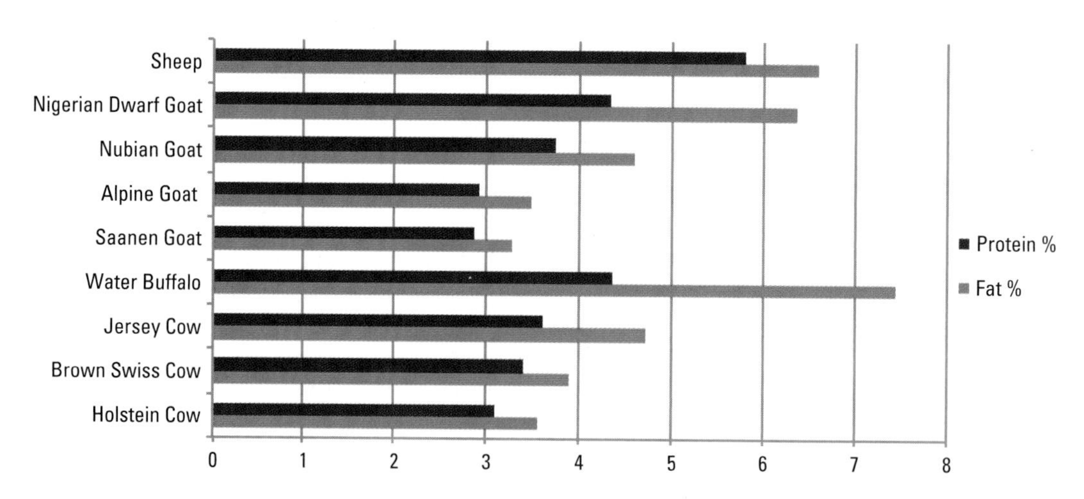

Chart showing some breed and species milk component differences. Keep in mind that usually the higher the components, the lower the milk volume compared to other animal breeds in the same species.

Sources: Data from "2010 California DHIA Report," the University of Illinois Lactation Biology website, and Yves M. Berger, "Breeds of Sheep for Commercial Milk Production," (Spooner Agricultural Research Station, University of Wisconsin–Madison)

days). Volume during that time is measured in pounds, not gallons. Official records collected through the Dairy Herd Improvement program from farms throughout the United States (more on that in chapter 8) show Jersey cows producing around 16,000 pounds per lactation and Holstein cows about 21,000. Dairy goat data from the same program shows standard-size does producing an average of 1,500 pounds of milk per year.

Milk Flavor and Aesthetics

Although all milk may be about the same color, it is far from the same in composition, flavor, and palatability. Some of these differences can be attributed to how milk is handled during and after collection (we'll cover this in detail in the next section of the book), and some of the nuances are due to the innate uniqueness in the milk from different species. If at all possible, you should try to taste the milk from all of the species and from several different breeds within those groups of animals you are considering as milkers. When you do your tasting, I hope you will have already read the next section of this book and will know how milk should be properly collected so you can determine if flavor issues are due to the animal itself or the fault of the producer.

Goat's milk, unfortunately, is still trying to dig itself out from years of bad press because of poor processing and unfortunate encounters. When I was first making the switch from wanting a cow to contemplating a milk goat, I made the mistake of purchasing some goat's milk from a well-known, high-end grocery chain. Due to no fault of this lovely store, the goat's milk was so bad our entire family dove for the sink to spit it out and rinse our mouths. Other goat people kept telling us that our own goat's milk would *not* taste like a musky, rutty male goat, but it wasn't until we tried that first chilled milk that I believed them. It was indistinguishable, to my palate, from whole cow's milk.

MILK MEASUREMENTS

When you begin shopping for dairy animals here in the United States, you will encounter several methods by which farmers and producers measure milk production. When milk is measured by volume—that is, gallons or cups—it will measure differently than if an actual weight is taken, because of the varying mass of the components in milk; remember that fat is lighter than protein, one of the reasons it floats to the top of milk. For a frame of reference, water typically weighs 8 pounds per gallon. A gallon consists of 16 cups. Each cup of water weighs 8 ounces or ½ pound. Nice and simple, right? Milk, however, typically weighs between 8.5 and 8.6 pounds per gallon. So 2 cups of milk weighs about 1.1 pounds. The doe I mentioned in the main text that produces 1,500 pounds of milk will have given between 176 and 178 gallons, and the cow producing 16,000 pounds will have provided 1,882 to 1,905 gallons of milk. I wish I could provide a simple rule of thumb for translating milk measurements for you!

Milk-tasting sessions provide an opportunity to train your palate to detect flaws and qualities of different milks. This session was held during a dairy class at Pholia Farm.

Goat's milk is also something of a paradox: The qualities that make it easier to digest also make it easier to damage. Mass-produced goat's milk is often doomed to a lesser state of sensory quality. The fact that goat cream does not readily separate from the rest of the milk can be a boon for those wanting to drink whole milk with an even, luscious mouthfeel, but a bane for those hoping to make butter.

Sheep's milk, which also has cream that does not separate easily, is preferred by many for its high protein, high fat, and rich quality. Without care, though, the distinctive taste of lanolin, a waxy substance secreted by wool sheep (some sheep not usually used for dairying actually do not have wool—or lanolin—but instead have hair) can enter the milk and give it a tacky, woolly flavor and mouthfeel.

Cow's milk has a mild taste, the degree of which can reflect (as can other species' milk) the breed, the season, and what the animals are being fed. The more yellow color of cow's milk, anything from light cream to golden, is due to the presence of beta-carotene in the milk (unlike goat's and sheep's milk, which contain vitamin A instead of its precursor beta-carotene, which is converted to vitamin A by our bodies). It is hard to resist the color of golden milk (so much more intense when the cows are dining on fresh grasses). Since cow's milk readily separates, or creams, it offers possibilities for flavor and textural changes thanks to the opportunity to remove some of the cream. All in all, cow's milk is likely to be less challenging in its flavor for most milk drinkers.

Land

Each dairy species is better suited to a specific type of land, including topography, climate, and native feeds, because of the environments in which they evolved and have been raised since domestication. In addition to these evolutionary factors, the amount of area available for livestock as well as proximity to neighbors should be considered when species shopping.

In general cows are flatlanders, goats are mountain folk, and sheep are somewhere in between. When necessary, goats adapt more readily to flat topography than cows will to steep terrain. And not surprisingly, the heavy weight of cows in comparison to sheep and goat can wreak havoc on slopes and hillsides where the soil is not well protected by deep-rooted grasses, terracing, or other soil retention qualities. If you plan on utilizing your land for grazing, foraging, or simple exercise, you'll need to take these things into consideration. If you will be keeping the animals in a more confined paddock, then topography is not an issue.

Land that provides year-round grazing is often also relatively flat. This type of feed and the damp weather that helps support it are more conducive to raising healthy cows than goats; cows eat with their heads lowered, cropping grasses, whereas goats evolved to eat plants such as shrubs and bushes, a type of grazing called browsing. Sheep seem to do well in either setting, being built for grazing grasses, as are cows, but also nimble and light-footed enough to navigate rocky and uneven terrain. In addition, those same wet climates that support grasslands are less than desirable for managing goats. Most dairy goats find rain as detestable as cows find a swarm of hungry flies and will not go out to graze, unless extremely hungry, when rain is falling. In the same vein, goats are particularly prone to parasite overload because of the larger numbers that are present on wet grasses and ingested during grazing.

Having a small farm on small acreage can bring you a lot of attention, some of it welcome, some of it not. If your land is situated next to a public throughway, such as a road or bike path, you can count on people wanting to pet, feed, and photograph the animals. You can also count on people making judgments about the way your animals are being managed. If there are residences near your farm, the odors and sounds of farm life may become an issue that strains neighborly relationships. It is a good idea to consider how life on the farm will look across a spectrum of seasons, from noisy females wanting to be bred, muddy paddocks and animals during rainy months, and flies and pests drawn to compost and barns to the barking of livestock guardian dogs and odors from manure and animals. Although there is no way to completely anticipate the reactions of your neighbors, diplomacy often is the best approach: Explain to them what a typical small farm will look like over the year in a manner that makes them feel that their input is valued. Setting expectations at the beginning is more beneficial than having to give explanations later.

Personal Disposition

There are goat people, cow people, and sheep people, and it has nothing to do with the animal you were born under in the Chinese calendar. You may not know your best barnyard match until you spend a little time with each species, then with some different breeds of the same species. It is inaccurate to make distinct

This young Saanen doe at Tumalo Farms in Oregon escaped from her paddock and was quite successfully eluding pursuit and capture until my husband, Vern, pulled out the camera and squatted on the ground. The young goat could not resist checking him out and was then easily secured.

Photo courtesy of Vern Caldwell.

generalizations about the personality of each breed of cow, goat, or sheep. Much will depend on what each breeder has selectively bred for, if anything. Good and bad dispositions can be found in all breeds. It is possible to give some general guidelines with regard to the different species' behavior characteristics, however. Although personality of your dairy animals may seem like a secondary reason for choosing one over the other, in reality it is one of the most important. You will be spending a lot of tough hours working with the livestock. You should not only enjoy looking at them, you should also enjoy working with them.

Dairy goats are inquisitive and highly expressive; hence their reputation as troublesome. To be a goat person requires enjoying interacting with animals who are comfortable expressing their opinions. Some breeds are more stoic, such as the largest goat breed, the Saanen. Others, such as the Nubian, are known for being expressive to the extreme. Again, these characteristics can be nurtured or diminished through selective breeding. If you purchase your stock from commercial or serious hobby breeders—who regularly milks their animals—your odds increase of getting animals with a good work ethic.

Milk cows have a reputation for a more indifferent personality than do goats. They can be easier to work with in situations in which dairy workers might vary and new people are common. As with goats, the largest breed, the Holstein Friesian, is known for its more docile, worker mentality. The smallest of the most common breeds found in the United States, the Jersey, can be quite the opposite. I'll never forget my first Jersey cow, Daffodil, and the obvious grudge she held for several days after I got a new cow.

Dairy sheep breeds in the United States are far more limited, with many dairy farmers crossing the most familiar variety, the East Friesian, with other, less historically productive breeds. When hand-raised, most sheep are petlike and docile. They are not quite as emotionally needy or as challenging as goats.

Starter Stock

I'll cover more details about healthy animals in the next part of the book, but for now I want you to understand the importance of finding a reliable source for your first animals. No matter what species or breed you want, if you cannot locate an honest, accomplished breeder, I would encourage you to choose another type. Besides educating yourself about the health issues for each species, you must be able to trust what the seller of your animals tells you and believe that their program has been one that encourages both good genetics and overall health. Remember that anytime someone is selling something they can become short sighted and focus more on making the sale than on developing long-term relationships with the buyer and maintaining their own reputation. Although auctions are a place to find very inexpensive livestock, they also can lead to expensive mistakes, with diseases and issues brought to the farm (in chapter 6 I'll cover policies regarding herd biosecurity). Wherever you purchase your stock, consider a quarantine period, and consult a veterinarian regarding what tests can be done to reduce the risk of introducing illness and unwanted pathogens to your herd and farm.

Surplus Stock

No matter how you manage your small dairy, there will be animals that must leave the farm. From excess young animals to poor producers that cannot be supported through their senior years, you should have a plan for how to deal with surplus animals. If you have cows, there will be far fewer offspring, usually just one each pregnancy, but goats have litters, with multiples of three or four not uncommon. Sheep commonly have twins. If you are growing your herd size or needing replacements, keeping young females is a good solution to offspring placement, but once the herd population is at its goal, a new management strategy must come into play.

Excess females with good genetics for milking can be sold, sometimes for a prime price. Males, usually 50 percent of the baby population, are not needed in the same numbers. I encourage people to face the reality of developing an exit plan for those animals. Options include raising them for meat, selling them to someone else for meat, selling them as pets (not likely a workable long-term plan), and dispatching them at birth (something more commonly done on large, commercial goat dairies). Females that fail to breed, are poor producers, have poor working dispositions, or are past their prime also must be managed. Sometimes these can be sold to people seeking just one or two animals for a home milk supply, but try to be honest about their failings; otherwise you will be creating an unhappy situation for both the buyer and the animal. They also can enter the meat supply and provide nourishment in a different fashion.

· 3 ·

The Business of Making Milk

Now that you have decided that a small dairy is for you, you might be entertaining thoughts of sharing your harvest of milk with others. If you can see yourself now, or one day, producing milk with the intention of selling, or otherwise distributing, it to farm-fresh milk enthusiasts, this chapter will help you through the steps involved in becoming a small-scale dairy producer. We'll cover such topics as options for marketing your milk, how to set prices, regulatory issues, and insurance and liability. In addition, I will be encouraging you to assess the viability of going into the small-dairy business by taking an honest look at the many costs and issues that must be addressed. Keep in mind that there is no "one size fits all" cost and income analysis that I can provide—the savvy business-minded farmer must do a good deal of homework before concluding whether milk production should be done as a hobby or as a living.

Market Options for Raw Milk

There are several options, in most states, for the sale of raw milk (see table 3.1 for a summary). At the time of finishing up this book, several states, including California and Illinois, had active groups making inroads into the legal acceptance of several tiers of raw-milk production in their states. Although not official yet, it bodes well for the future of those considering entering the market or wanting to add raw milk to their product lists. Since what is acceptable now may be changing—for good or bad—be sure to investigate what is legal in your state. The bibliography has contact information for the Weston A. Price Foundation's Campaign for Real Milk website, where such information is up to date. Some of the laws are particularly convoluted and difficult to understand, but you can contact a Real Milk representative for clarification.

In general there are the main options (with some creative hybrids) for getting raw milk into the hands of the consumer: sales in retail settings such as grocery stores; directly from the farm, with the consumer picking up their share; through a community supported agriculture (CSA) approach such as a herdshare, cowshare, or lease; and in bulk to a cheesemaking facility. Let's go over these options, with the assumption that they are all legal in your state.

For raw milk to be sold in a retail setting, such as a health food store, grocery store, at some categories of farm store, or at most farmers' markets, the milk must be produced in a facility that meets licensing requirements and is currently licensed to sell such a product. A retail sale is usually defined by the availability of a product to any customer who pays for it at the place of sale. This is important to understand, since some transactions fall outside this definition. I'll talk about those in a bit.

These appealing stickers are produced by the Farm-to-Consumer Legal Defense Fund, one of the most powerful voices in advocating for the rights of small farmers, artisanal producers, and consumers seeking to source food directly without regulatory interference.

Milk sold at the retail level by the farmer, such as in a farm store or farmers' market, is sometimes in a different category from other retail venue sales. When the producer sells the milk to a store that then resells it, that is referred to as "full retail." Full retail sales of raw milk are the least common option in the United States, with only a few states allowing this sort of commerce. The producer selling to a retail store will sell the milk at a "wholesale" price, anywhere from half of the full retail price. Although producers obviously makes less income when selling wholesale, they also have less overhead and associated costs. They also have a direct link with the consumer, which adds value for both the producer and the buyer, with a definite, articulated link to the land and animals.

WHAT ABOUT SELLING MILK AS PET FOOD?

Several states allow the sale of raw milk "for pet" or "for animal" consumption. In some of these situations, you are required to decharacterize the milk by adding an approved food coloring or charcoal. In some you are required to obtain a feed permit for the production and sale of animal feed. In some cases, however, the label is really a wink-wink kind of deal, where both parties know that the milk is going to be consumed by two-legged animals. On the other hand, there are small-dairy farmers who really want to sell their milk for pet and animal use. Raw milk, especially from goats, is highly valued for raising puppies, orphaned deer, and other livestock. Producers rightly worry that buyers proclaiming it is for animals might in truth be drinking it themselves. These dairy people often do not want the added labor and responsibility involved in collecting milk of a quality fit for human consumption. Even when not required, I would recommend that if the end user is truly meant to be nonhuman, then decharacterizing the milk might be a good idea.

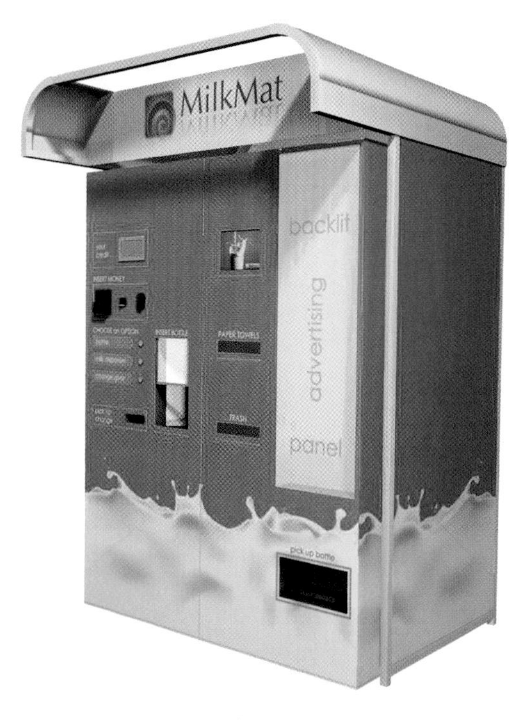

Raw-milk-dispensing machines can be found in several European countries. The machines typically dispense milk from only one farm and provide a novel way for farmers to sell their milk directly to the public.
Photo courtesy of Konrad Pszowski, MilkMat, S.C.

Probably the most commonly allowed way to sell raw milk is directly from the farm, often at a farm store or a self-serve stand. If your farm is located convenient to a public road as well as a population of potential buyers that regularly pass by, then a farm stand might be a fantastic opportunity. Be sure to find out if any local ordinances prohibit such an activity, but in most states "right to farm" laws support the right of farmers to sell their own products directly from the farm. Of course, you also must make sure that your product meets any laws that define how it is produced.

Dairies that sell milk directly from the farm handle the details in different ways beyond a simple farmstand. Some farms allow customers to fill their own milk jugs, or they fill them, directly from the milk bulk tank. Producers that do this are sometimes referred to as "juggers." Others fill containers that are either owned by the customer or are loaned by the farm, or a bottle deposit is collected. How bottles and jars are dealt

Whole, raw cow's and goat's milk for sale at an on-farm honor stand at Runnymede Farm, Oregon.

TABLE 3.1 The Legality of Raw-Milk Sales and Distribution by State (as of September 2013)		
Category	**States**	**Comments**
Retail sales legal These states allow sales in retail stores and require a permit that also allows for sales on/off farm and at farmers' markets.	AZ ME PA CA **NH** SC CT NM WA **ID** **OR***	* goat & sheep milk only
Licensed on-farm sales legal	MA SD[4,6†] WI **MO**[4,6] TX NY UT[†]	† sales legal at retail store if producer owns and operates the store ‡ sales legal at retail store if producer has majority ownership in store
Unlicensed on-farm sales legal	AR[3] MN NE IL MS[1,2] **NH**[3,4,6] KS **MO**[4] OK[3†]	§ limit on volume of goat milk sales
Herdshares/Cowshares	AK[b] MI[d] TN[a,8] CO[a**] ND[a] WY[b] **ID**[a**] OH[c]	** Farms operating share programs must register with the state.
Sale of raw milk for human consumption illegal	AL[7,8] IA NJ DE KY[2,5,7] NC[8] FL[8] LA RI[2,5] GA[8] MD VA[7] HI MT WV IN[7,8] NV[††]	†† requires County Milk Commission (CMC) approval; only Nye County has a CMC but has not approved any dairies yet

Source: *Copyright 2010-2013 FtCLDF, www.farmtoconsumer.org and used with the permission of Pete Kennedy, Esq, president, FtCLDF*
Note: *Bolded states appear in more than one category.*
[1] *Limit number of lactating animals*
[2] *Only sale of goat milk legal*
[3] *Limit on sales volume*
[4] *Delivery legal*
[5] *By doctor's prescription only*
[6] *Sale at farmers' markets legal*
[7] *There is no law either expressly legalizing or prohibiting herd shares; state is aware herd share programs currently exist and has taken no action to try to stop them*
[8] *Sale of raw pet milk legal*
[a] *Legal by statute*
[b] *Regulation*
[c] *Court decision*
[d] *Written policy*

with varies, from customers to the producer being responsible for making sure they are clean and sanitized. Plastic jugs are typically not returned or reused by the milk producer.

A novel approach to getting milk into the hands of consumers is to enter into a partnership of sorts that includes selling a percentage or portion of ownership in a single animal or the entire herd. Typically the part owner pays a periodic fee to support the animal's care. This ownership, in theory, allows for any products produced by the animal to be consumed by the owner. I say "in theory," since some states' or jurisdictions' commerce laws have made this sort of arrangement illegal, difficult, or one that requires regulatory oversight. Characteristics that typically describe a proper herdshare arrangement include these:

- A signed ownership agreement between the farm and the buyer stipulating the buyer's percentage of ownership in an individual animal or herd, entitling them to a given percentage of milk produced.

- A signed agreement setting the monthly or other regular payment that covers feed, care, and management related to the percent of ownership.
- Whether milk is to be picked up at the farm, delivered to the owner, or picked up at a disbursement point such as a farmers market.
- That milk will never be sold at a distribution point; in other words, if a new buyer arrives, they must purchase a share, then come back to collect milk produced after the buy-in occurs.

Neither a herdshare nor a direct retail sale to the end customer, raw-milk buying clubs are also common in some parts of the country. A buying club typically utilizes a person, who is usually not the farmer, to act on behalf of the milk purchaser in delivering the product to the club members. There may be a charge for the delivery service. Supporters of buying clubs argue that the person delivering the milk is not selling the product but is simply offering a delivery service. Many of these folks have crossed paths with the law, especially when delivery includes taking milk across state lines. Rules established in the late 1980s prohibited the sale of raw milk outside the state in which it was produced and regulated. Even pasteurized milk producers must obey different regulations when milk is to be shipped across state borders. A person can in theory still buy a gallon of legitimately produced raw milk and drive it to his home in another state, but in most cases, any transfer of ownership will be seen as commerce and, as such, is regulated.

Selling raw milk directly to a licensed cheesemaking facility can be a great option for those wanting to add income to their farms but not wanting to deal with packaging and marketing their product directly to the consumer. Small artisan cheesemakers who don't have their own herds of cows, sheep, or goats often have a difficult time obtaining high-quality raw milk for the production of handcrafted cheese. With cheesemaking on the rise, the demand for milk

This dairy farmer cut infrastructure costs by using repurposed shipping containers for milk processing rooms.

SMALL DAIRY PROFILE: WILLOW-WITT RANCH, OREGON. GOAT'S MILK HERDSHARE.

When you hear the word "ranch," you might picture sprawling vistas of sagebrush with rangy beef cattle dotting the landscape, or you might picture the Ponderosa, of 1970s TV fame. The 400-some acres of high mountain land farmed by Suzanne Willow and Lanita Witt is more picturesque than either. With natural alpine meadows, wetlands (being restored by the pair), and craggy pines and fir, Willow-Witt Ranch is breathtaking. Lanita, a practicing OB-GYN doctor, and Suzanne, formerly a physician's assistant, tend their off-grid farm with a caring philosophy of sustainability, biodiversity, and multiuse. In addition to their dozen or so Alpine goats that supply raw milk to a number of people through a carefully managed herdshare program, Willow-Witt produces and sells inspected meat products and eggs; features an adorable farm-stay apartment and a multiuse campground complete with two "glamping" (a term implying glamorous camping) tents; and offers internship opportunities.

Suzanne says they originally had a couple of goats to supply milk for their family and later to raise piglets for 4-H projects. People buying the young hogs and other farm products began pestering them about possibly buying milk. For more than a decade Willow-Witt made its milk available directly from its farm store under Oregon's state laws allowing for the legal sale of the milk from nine or fewer goats. This worked well, but because of the remote location, it was difficult for many customers to make the trip consistently to the ranch, so sales would fluctuate drastically at times. Oregon, at this time, has no laws restricting herdshares, so the couple went about setting up such an approach to making their milk available.

By calculating the average amount of milk produced by each of their goats for a full year (they milk their Alpines year-round), as well as costs for the organic feed they buy, cleaning supplies, and labor, they settled on a monthly fee of $30.00 to buy one share of the herd. (Suzanne said that feed costs and infrastructure improvements have made that fee a bit too low.) The amount of milk each shareholder receives varies somewhat but seems to average about ½ gallon per week, Suzanne told me over the phone. She said some people have purchased multiple shares while others have invested in only a half share. Willow-Witt utilized the resources of the Farm-to-Consumer Foundation (FTCF) to develop its contract. (See resource section for more on FTCF.) When a new member buys into the herd, there is also a small fee for executing the contract. Suzanne says they buy back the share (minus a percentage) when a person decides to leave the plan.

Since both Lanita and Suzanne have extensive medical training and experience, food safety and animal health are imperative to them. Their herd is tested and is free of diseases that could pass to humans as well as those that harm the health of the animals. The milk is rapidly chilled during milking, and bottled milk always is stored at verified refrigeration temperatures. To help educate the shareholders, the practical booklet *Safe Handling* by Peg Beals is included in the initial membership. The milk is regularly tested on the farm (as of this year) using some of the testing methods described in chapter 10.

For Willow-Witt Ranch the sales of raw milk represent both a way to support the biodiversity of their farm, including the forest maintenance work that rotational browsing of the goats allows and a way to supply a product that is in great demand, and in almost nonexistent supply, in their part of the country. They manage their dairy with vigilance and education and are well worthy of the title "Raw-Milk Role Model." For more about the farm's programs visit http://www.willowwittranch.com/.

produced by small dairy producers is also increasing. To be used by a commercial producer of any size, the milk source also must be licensed, be inspected, and meet all required regulations.

Typically, milk is picked up by the cheesemaker, but it also might be delivered in milk cans or in a small milk tank to the cheesemaking facility. The milk producer will receive far less money for the product when sold this way, but will have fewer costs, less labor, and fewer business management issues. No matter what the volume of milk is, I highly recommend entering into a contract that protects both parties and defines each side's expectations and duties (a sample milk purchase agreement appears in appendix C).

What's It Going to Cost?

It is impossible for me to give you an accurate idea of what your costs will be to build and operate a small dairy. There are far too many variables, including your own expectations of what constitutes a "good living." In chapter 2 I introduced you to the concept of the small farm as a lifestyle business, one that in the best cases supports the farm lifestyle but rarely provides any type of investment or measurable profit. To decide if a small dairy is likely to at least pay for itself, you will have to do your homework, gather the numbers, then do the math. To help you toward that, I have included several charts and forms that address many important considerations related to finances, labor, and personal suitability.

Water and wastewater management is often heavily regulated. This twenty-goat dairy was required to capture roof runoff from their small barn and divert it to tubs and other parts of the property.

Table 3.2 will help you estimate or document the spread of labor throughout the year. Record the average hours per day (estimate or average), then the totals. This will help both in figuring out how to manage your own time and in anticipating labor needs with changes in herd size, season, and so on. It is always a good idea to overestimate how much time you will spend doing these tasks. Farmers whom I know (including ourselves) are typically behind on many projects, repairs, and needed work, especially when money is required to move forward. This is not a failing; it is a reality of farming—one that is often difficult to accept.

Table 3.3 should help you estimate income and ongoing expenses. Again, don't use "hopeful" numbers; be as realistic as possible. Typically, things cost more than you expect, and I have yet to see feed, fuel, and utility costs go down!

There are several in-depth publications that utilize studies, usually done through university Extension program grants or funds, that look at data in varying parts of the United States in regard to costs and income related to starting a commercial dairy. You also may be able to access more regional information from your state's agricultural land grant university Extension service. Should you utilize these types of publications, you will find an incredible amount of information and data, most

A Flow Chart to Help You Determine Your Suitability for Running a Small Commercial Dairy

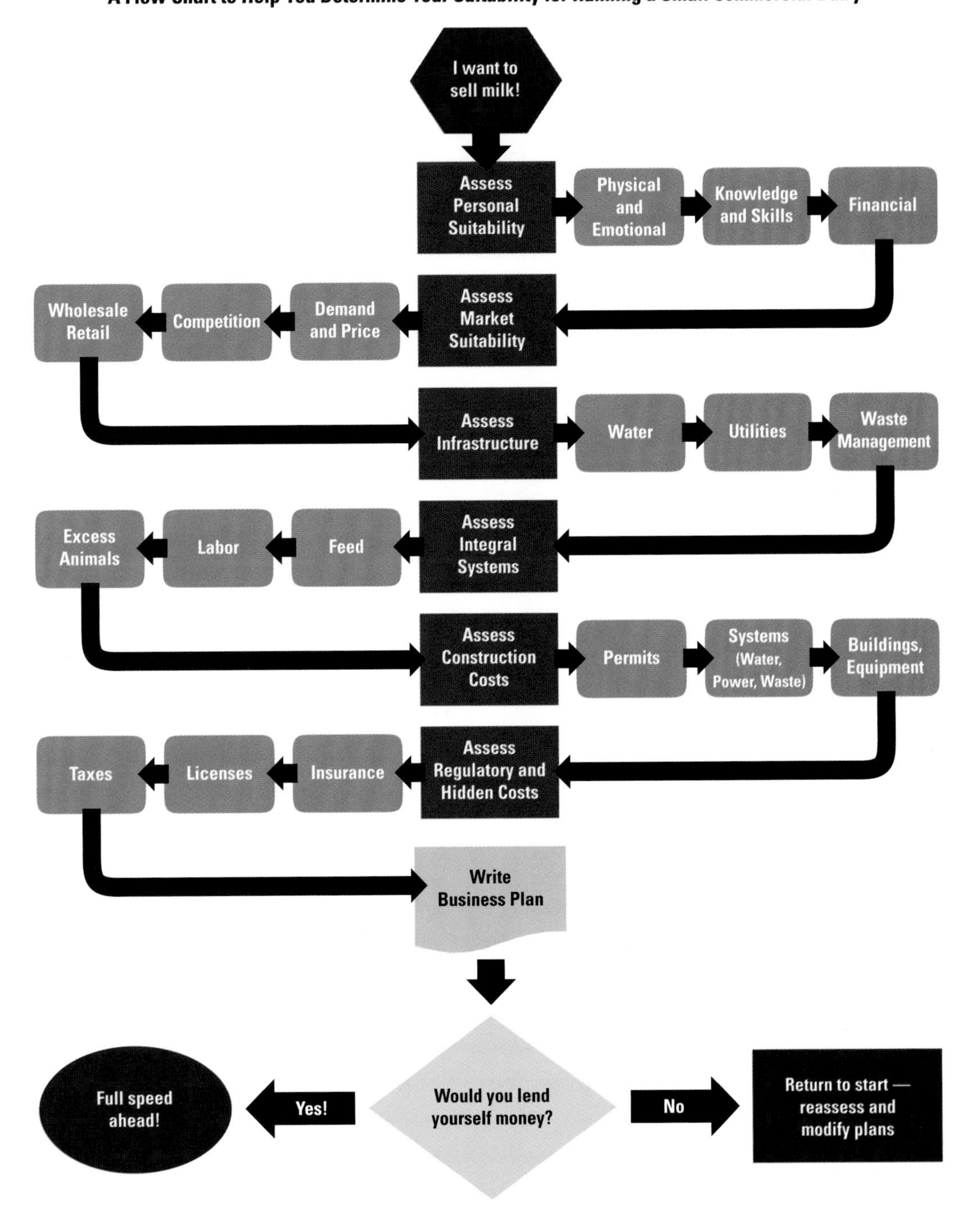

TABLE 3.2. Labor Estimates/Log

	January–March Hours per Day/Total	April–June Hours per Day/Total	July–September Hours per Day/Total	Hours per Day/ Total
Maternity Care and Breeding				
Young-Animal Feeding and Bottle Washing				
Young-Animal Pen Cleaning				
Young-Animal Medical and Procedures				
Young-Animal Milk Processing				
Record Keeping and Sales Time				
Milking (include extra time training new animals)				
Cleanup from Milking				
Pen Cleaning				
Feeding				
Hauling Feed and Bedding				
Procedures and Medical Care				
Equipment and Facilities Maintenance, Repairs				
Milk Bottling and Cleanup				
Delivery of Milk and/or Sales Time and Record Keeping				
Website, Phone Calls, e-mails, Other Business				

of it quite helpful. As I said earlier, the most accurate way to project earnings is to do your own in-depth study. But in the least, take a look at the work that others have done to help you through the process.

Pricing Boutique Milk

Whether it is pasteurized or raw, cream-top cow's milk, or snowy white goat's milk, small-batch locally produced milk has fantastic niche marketing possibilities. Even large grocery stores with shelves lined with gallon after gallon of commodity, highly processed milk will likely have a natural foods section showcasing other dairy options. I mentioned in the last chapter that raw-milk prices have a fairly uniform range across the country. That being said, prices also depend on demand and availability, not unlike any other product. The degree of oversight, such as whether the processor is licensed and inspected or not, does not often impact the price, which might seem unfortunate, given the costs involved in building and maintaining a Grade A, or even a Grade B, facility. Typically, goat's milk commands a bit higher price than cow's milk. Sheep's milk is rarely available at the consumer level, but it brings the highest price to those selling to cheesemakers.

TABLE 3.3. Ongoing Expenses and Income

Item	Frequency (some costs might be annualy, such as feed or insurance and need to be anticipated)	Estimate or Actual
Units of Milk (number of animals × herd average production ÷ unit size × price)		
Sales of Live Animals or Meat		
Feed Costs: Forages (typically annual purchase)		
Feed Costs: Grain and Supplements		
Bedding Costs		
Veterinary or Medical Supplies		
Breeding or Replacement Costs		
Labor (include your own)		
Equipment Repair and Replacement (include fencing, tractors, vehicles, dairy equipment)		
Utilities		
Chemicals		
Bottling Supplies		
Website and Phone		
Insurance		
Property and Business Taxes		
License and Testing Fees		
Membership Dues (typically once a year)		

These raw-milk single-serving containers, produced by Amish farmers in Pennsylvania, were being sold at a farming event.

When sold at the farm, raw cow's milk ranges from $6 to $12 per gallon, depending on location. Raw goat's milk averages $8 to $14 per gallon. Where legal retail sales are possible, prices can come closer to $20 per gallon, depending on the location. If sold in smaller-size containers, such as half gallons or quarts, prices might total higher per gallon. For example, single-serving sizes might cost $3 each, which would come to $24 per gallon. But the same farmer might sell gallon-size containers for $12. These size and price choices are important to consider as you develop your plan and analyze the market potential of your area and its clientele.

Raw cow's milk sold in bulk to a small cheesemaker ranges in price from $1.50 to $3.00 per gallon and goat's milk from $2.50 to $5.00 per gallon. Much of the pricing surrounding bulk sales is dictated by milk quality, including butterfat and protein levels, somatic cell count, and bacteria counts. Moderate-size cheesemakers also might pay incentives for producers willing to continue providing milk through the winter. A premium price also is paid for humane and organic certified milk.

If small-production milk is available in your area, whether it is raw or pasteurized, you can do some initial market research quite easily by collecting this information. As I mentioned earlier, prices tend to be much higher where demand is high but supply is low—this is especially common in areas near a large, progressive metropolitan population.

Regulatory Issues

In most cases you will encounter several layers of licensing and regulatory approval before you can begin to sell milk. To obtain a dairy producer's license, your facility must meet approval and your milk quality also must meet certain criteria. The initial license is maintained by paying an annual fee as well as continuing to meet the criteria of the state. The criteria not only vary greatly by state but are also subject to change. Even as I write this, new federal regulations for all food producers are in the works thanks to the Food Safety Modernization Act (FSMA) and will create additional burdens for many producers. If you become involved in the sale of products from your small dairy, I encourage you also to become involved in the support of laws protecting food rights.

When you begin looking into producing milk for sale, I don't recommend getting licensing and regulatory information only from another farmer, since there may have been changes in the laws or confusion on the part of the farmer—even if he or she is currently licensed. This is especially true in states that have multiple categories of allowed sales. Dealing directly with the regulatory body that oversees state or regional laws is your best bet for finding the correct and current information. Typically, this will be the state's department of agriculture. But in some states it may go by a different name or be superseded by a more local law. This is the case in California, where the retail sale of raw milk is legal statewide but is prohibited in several counties.

Some jurisdictions offer tremendous support and guidelines for producers seeking to sell raw milk, while others—even where it is legal—seem reluctant to support and guide producers. In most cases inspectors are fairly objective and able to separate their own opinions from the regulations. When dealing with officials, an attitude of humility and respect for the regulations is helpful in gaining information and establishing a professional relationship. Here are some good initial questions for the inspector or other pertinent party:

- Are raw-milk sales legal at any level in my area?
- Can you direct me to or send me the statutes that define the sale of raw milk in my area?
- Does my area define and/or control the distribution of products to owners who do not live on the farm (a.k.a. herdshares)?
- If my facility will need to be inspected, at what point may I invite you out to see the plans or progress?
- Are there any publications available or that you recommend that tell me what exactly will be required for my milk sales to be legal?

TYPICAL STEPS TO LICENSURE:

1. Contact your state regulatory agency in charge of inspecting and licensing dairy plants.
2. Ask questions from the list on this page regarding regulations, inspections, and guidelines.
3. Have facility inspected, labeling approved, animals tested (where required), and milk tested (where required). (See appendix B for sample inspection forms.)
4. Pay annual fee to maintain license.
5. Make facility available as required for inspections.

I recommend turning to peer groups for such things as educational opportunities; lobbying and political action; pooling resources (such as for purchasing bottles and lab supplies that might be sold in numbers too large for one small producer); and camaraderie. Although not all states and areas have small dairy producer groups, you can interact with many producers through social media and Internet chat groups. Attending meetings of larger organizations such as the Weston A. Price Foundation—which has regional chapters throughout the United States—also will allow you to connect with other concerned and involved individuals.

Insurance and Liability

If you have a farm—of any kind—and other people participate in anything related to the farm, such as visiting or using its products, you might become responsible for any mishaps that occur. There are several major law firms actively seeking clients for whom raw milk might be implicated as a source of illness. These firms and attorneys have taken a passionate, aggressive stand. Even where the product is legal, the right situation and the right lawyer could make your life very difficult. Although this entire book is dedicated to providing knowledge and guidelines to

THE ACCIDENTAL CONSUMER

If you drink raw milk, consider it safe, and regularly have it in your refrigerator, it can be easy to forget that your preferred beverage is special. A 2012 outbreak of *E. coli* 0157:H7 from raw cow's milk provided to over forty families in Oregon included one young victim who unknowingly drank the milk while playing at the home of one of the herdshare members. You can imagine the added outrage on the part of this sick child's parents. At no point were they a part of the decision-making process. I would encourage any family that chooses raw milk to train all family members to treat the milk as restricted to their use only. No matter how much you believe in consuming the product, it is still a decision that not everyone would make.

help prevent problems, chapter 11 will help you create a sound food risk reduction plan as a means to help prevent such an unwanted situation.

As producers of raw-milk cheese, we have encountered difficulties finding a complete policy. Our current coverage is provided by two carriers. The first provides the usual property and auto coverage. The second carrier provides our product liability coverage that is supposed to protect us in case our products are ever implicated in an illness. We used to have a single carrier, but after a few cases of raw-milk cheese-related illnesses in a neighboring state, carriers in all parts of the country started dropping all producers of farmstead raw-milk cheese. Even in states where the product is legal and regulated (meaning inspected), some insurance companies refuse to cover foods they categorize as high risk.

My best advice for finding the right carrier and policy is to find a superior insurance agent—one who represents others in the same field of work. It may mean leaving a company that you have been with for years and feel loyal to. But you need to work with someone who knows the ins and outs of your industry.

The cost of product liability insurance will be based on several factors. The perceived risk of the product, the dollar value of your sales, the scope of your sales—local, regional, and so forth—and things you do to limit risk, such as continuing education, food risk reduction plan, and security measures. When you get to chapter 11 and the chore of creating a food safety plan, keep in mind that the effort you expend might not only save you money but might make the difference between obtaining coverage or not. While I hope you will never need the protection of an insurance policy, you should be aware of where you would stand legally should an issue arise.

Business Plan

If you have been doing your homework and keeping track of data such as costs of feed and prices you can charge for milk, you are probably ready to write your business plan. Although you may think that a business plan is needed only if you intend to seek a loan, any business worthy of your time and investment deserves a good business plan.

The complexity of a business plan can reflect the complexity of the business—the simpler the company and product line, the simpler the plan. There is no single best template for any small business, but a good business plan should include information that gives a full overview of the company. Here are some typical important parts of a solid business plan:

- **Company Overview or Executive Summary:** This section appears first but should be written last; after you finish the more detailed sections, it will be easier to write a brief introductory summary of the entire plan. Think of this section as what you would tell a bank manager if you are given a five-minute interview. If you will be seeking funding, this section should include how much money you are requesting.
- **Company Description:** This expansive section will include many subsections, including the company mission, goals and objectives, business philosophy, who will buy your products, an overview of your industry, and your business structure (such as sole proprietor or LLC).
- **Product Information:** Describe your product, its packaging, its pricing, and its qualities.
- **The Market:** In this section describe the market for your product, including market research and data that supports your statements. Describe how your product will fit within this market and be desirable to customers.
- **Operational Plan:** Describe how your product will be produced; include quality control, customer service, and inventory control. Next, discuss what type of facilities will be needed or used, including issues to do with their maintenance and costs. Issues to do with personnel also are covered in this section, including how many are employed and worker safety and training. Finally, discuss legal issues such as permits, licensing, and insurance.
- **Financials:** This is perhaps one of the most difficult sections, since you hope to create a budget that includes estimates of income and expenses for the first few years of your business. If you're seeking funding, you also will need to list personal assets and capital that each owner possesses. If you're not seeking funding, this section is still important, since it will help you assess whether your idea is financially realistic.
- **Exit Strategy:** Unlike many careers, a home dairy cannot be "quit" or retired from easily. You should have a plan for how you will disburse assets and deal with liabilities should the company need to cease doing business.

If you will be seeking a loan or other investment for your dairy, a superior business plan is essential. Fortunately, help is available through small business associations and online. Classes are also commonly available through local colleges and business development organizations. If your business is to be more than just a hobby, I cannot express the importance of creating a well-thought-out business plan. It may not turn out to be as accurate as anticipated, but it still will help you envision your future—and may even save you some costly mistakes and regrets.

PART II

THE PHILOSOPHY, SCIENCE, AND ART OF THE SMALL DAIRY

·4·

Microbiology and Milk: Understanding the Basics

Before you can fully understand the management methods that will produce top quality milk, it's imperative to appreciate the microscopic world and how it relates to this amazing liquid. In this chapter, I will cover some basics about bacteria including why we literally cannot live without them, the varieties that milk is most likely to attract, and how best to protect milk from harmful bacteria. Even if you don't have a strong science background, this quick study of microbes will give you the perspective, respect, and knowledge you need in order to make the proper animal and milk management decisions.

Bacteria: We Can't Live without Them

In chapter 1 I mentioned the advent of the germ theory of disease in the late 1800s and the subsequent fear of and warfare with microbes that is still ongoing. The consequences of this century-long battle fought with sanitizers, antibacterials, antibiotics, and germaphobia are profound—from bacterial resistance to medications and weakened and malfunctioning immune systems to new strains of more powerful pathogens. Fortunately, a new paradigm is developing—one in which people are also becoming aware of the important positive role microbes play in human life and health. In fact, according to an article in *Scientific American*, a healthy person hosts ten times more bacteria than our own biological cells. Even though they are much smaller and less complex than our own cells, they still account for about 3 pounds per healthy person, according to Sandor Katz, author of *The Art of Fermentation*. It's hard to imagine the number of unseen, tiny organisms that it would take to make something that weighs just about as much as our own brain! And guess what, we would be dead without them. This massive community of microbes is known as the human microbiome.

In our bodies, and those of all members of the animal kingdom, bacteria reside on the surface of the largest organ we have—our skin. Our skin not only covers the outside of our bodies but also lines the inside of all of the openings and forms our digestive tract—a long tube that begins at our mouths and ends at the anus. While bacteria live on all skin surfaces, the majority reside in the lower part of our digestive tract.

Milk in the udder of an animal without an udder infection is sterile (we'll learn more about this in chapter 5). "Sterile" means free of living organisms, such as bacteria. It doesn't mean that it is "dead" with regard to nutrients, enzymes, and life. Indeed, a human baby in its mom's uterus that sucks its thumb, has a heartbeat, and is ready to be born is alive, but sterile. Guess where the little one gets its first dose of bacteria? During and immediately following a vaginal birth, babies begin collecting bacteria from everyone, and everything, they come in contact with. This exchange of bacteria continues throughout the lives of all creatures, with new bacteria being introduced through food, environment, and personal interactions. Although most of these bacteria are either harmless or beneficial, some are pathogenic. Later in this chapter I'll talk about some of the specific pathogens of concern for dairy producers.

Life under the Microscope

Organisms that cannot be seen without the use of a microscope are called, not surprisingly, microbes or microorganisms. Bacteria aren't the only microbes in this group; yeasts and molds (fungi), protozoa (single-cell animals), and viruses are also microbes. Although viruses are not living organisms, their study still falls within the science of microorganisms, or microbiology. For the most part milk producers and consumers need to know about bacteria, although if the milk is destined for cheesemaking, yeasts, molds, and viruses are important as well.

I find it helpful to think of microbes as existing in three main groups—pathogens, which can cause illness; spoilage microbes, which cause food to "go bad"; and those that are beneficial, including probiotic, acid-producing, and other helpful organisms, such as those used in cheesemaking. Because of the damage caused to people and their animals by food-borne pathogens and the cost to the food industry by spoilage microbes, it's not surprising that science puts the spotlight on these categories—especially in terms of research dollars and, consequently, published data and findings. With the increased appreciation of the human microbiome, however, the things that damage it, such as a diet high in processed foods, low in nutrients, and low in prebiotic and probiotic bacteria, are now also being researched in the scientific realm and are frequently the subject of discussion in the popular media.

The Right Conditions for Microscopic Growth

Bacteria, like all microbes, require the right conditions in order to complete their full life cycle. Each microbe has its own preferred conditions. These conditions include ideal **pH**, **moisture**, **temperature**, **oxygen**, **nutrients** (food for the bacteria), and the **absence of growth inhibitors** such as antibiotics and natural inhibitors (see sidebar for more on the natural inhibitors in raw milk). If the right

NATURAL ANTIMICROBIAL CHARACTERISTICS OF MILK

Milk is an amazing fluid designed by nature to provide safe nourishment to young mammals, and it contains some built-in protections that limit bacterial growth. These protections, however, are not intended to function for long—since milk is meant to be consumed frequently and directly from the mother. Antimicrobial factors in milk include the lactoperoxidase system—the enzyme peroxidase kills bacteria by oxidizing; lysozyme—an enzyme capable of damaging bacteria cell walls; and lactoferrin—a protein capable of binding iron, making it unavailable to bacteria that need it for growth.

The lactoperoxidase system is of particular interest since its activity level in milk can be sustained and extended through the addition of two compounds—thiocyanate and a source of peroxide. These two compounds effectively sustain the reactions that occur during this system and serve as extremely effective antimicrobials. They are so effective, in fact, that their use is well documented and supported by the World Health Organization (WHO) as a milk preservation method in parts of the world without refrigeration. When used properly, milk can be held at room temperatures for quite some time without any bacterial growth.

The topic of milk's natural germicidal or antibacterial properties, has been researched and debated for at least the last hundred years. Cheesemakers using raw milk can sometimes come up against the natural defenses in unpasteurized or heat-treated milk—causing the prevention of the proper growth of even good lactic acid bacteria. But rarely. In fact, some cheesemakers don't add any starter culture, instead allowing the native bacteria that get into the milk during milking (from the teat surface and environment) to grow and coagulate the milk. This tells us that these natural systems can be overrun easily—given time and the right conditions. Luckily for drinkers of raw milk, proper collection and storage should hold the milk in stasis and ensure reasonable levels of safety.

habitat doesn't exist, then bacteria—both "good" and "bad" bacteria—will not thrive and may even die.

Many things supply the right balance of conditions to support the life cycle of microbes. Living creatures provide the right environment for hosting a vast population of microorganisms. Anything that disrupts a person's normal, healthy, and balanced condition, called homeostasis, can make that person more susceptible to a change in the balance of beneficial bacteria—changes that might favor pathogenic varieties. Let's take a deeper look at these conditions so we can begin to appreciate and understand how best to protect milk, and ourselves, from the microbes we don't want.

The **pH scale** is a system for assigning a number value to substances based on how acidic or alkaline they are when compared to pure water—which is neutral. On this scale pure water is assigned the number 7.0. Values lower than 7.0 are increasingly acidic, while those above 7 grow increasingly alkaline. Fresh, clean milk is usually between 6.6 and 6.7, while blood is just above 7.0. Typically, the closer to neutral a substance is, the more appealing it is as a growth medium. When milk is turned into products such as cheese, yogurt, and buttermilk, it is acidified to much less pathogen-friendly pH levels of between about 4.6 and 5.2. This is the primary way that cheese is protected from spoiling and from growing harmful microbes. However, some unwanted microbes have the ability to grow resistant to acid, including the worst strains of *E. coli*.

Milk is made up of a lot of water, meaning that there is plenty of **moisture** to support the growth of microbes. The term that describes moisture content in this regard is water activity or "a_w." (I remember this by thinking of it as "available water.") Various processing techniques can reduce the water activity of milk and therefore the potential for microbial growth. These include dehydration of milk to a dry powder; transforming the milk into a dry, hard cheese; and the addition of salt (which essentially binds the water to itself and makes it unavailable to microbes). When the goal product is fresh, fluid milk, the potential for microbial growth in good part due to moisture content remains relatively high.

Bacteria can be grouped by temperature growth range. These ranges overlap, and few are exclusive to the growth of only "good" bugs. Bacteria that do well in cooler temperatures are called psychrotrophic. Many bacteria that cause food to spoil fall into this category. Although they don't typically grow well at refrigeration temperatures, anywhere from just above freezing to about 40°F (4°C), they do survive and will begin to grow when conditions are a bit warmer. Warmth-loving bacteria are called mesophilic (I like to remember this by thinking of "m" for middle temperatures). These types typically prefer temperatures to be right around the same as our bodies, in the upper 90s Fahrenheit (37°C). If bacteria do well in even higher temperatures, they are called thermophilic (think of a thermos to keep things warm).

When milk leaves the udder of the animal, it is at a wonderful temperature for mesophilic bacteria to begin to grow. As it it is cooled, but before reaching refrigeration temperatures, there is the opportunity for any psychrotrophic bacteria present to begin to multiply. So until milk reaches cold refrigeration temperatures, it is quite vulnerable to the growth of many different bacteria. When heated above about 120°F (49°C) bacterial growth will be inhibited and many microorganisms will be destroyed, depending upon the temperature reached and the length of time it is maintained. This is the reasoning behind heat treatments such as thermization (a technique for heating milk to temperatures below that of pasteurization to reduce bacteria populations) and pasteurization. For more on temperatures and times associated with heat treatments, see chapter 13. Freezing does not kill any of the categories of bacteria. Frozen milk will not experience bacterial growth while frozen, but any bacteria present will "wake up" once their growth temperature range is regained.

When you are attempting to save the nutrients in milk for consumption by humans, instead of by bacteria, it is necessary to deprive the microbes of any other conditions that will allow them to grow.

One of the main ways that scientists categorize bacteria is by their need, or lack of need, for oxygen. There are three categories that are used to differentiate these needs. First there are bacteria that can live and reproduce only in the presence of oxygen called obligate aerobes (think of aerobic exercise—which burns oxygen—and the word "obligated," which means you "have to do something"); next there are *obligate anaerobes*, which are just the opposite and thrive in oxygen free situations; finally there are those microbes that prefer oxygen but will adapt to a zero oxygen environment are called *facultative anaerobes.*

If you are starting to wonder what all these terms and definitions have to do with milk, let me tell you that some of the worst pathogenic bacteria are facultative anaerobes—this means that some of the nastiest bugs out there are also the most adaptive to different situations. Luckily, many beneficial bacteria are also facultative anaerobes and are responsible for the fermentation of milk (fermentation takes place without the need for oxygen) into tasty products such as cheese and yogurt.

Milk is filled with nutrients that are essential for humans and are also attractive to microbes. When the conditions just covered are optimal, bacteria can begin to process the nutrients specific to their needs. When you are attempting to save the nutrients for consumption by humans, instead of by bacteria, it is necessary to deprive the microbes of any other conditions that will allow them to grow and increase their own numbers and subsequently cause the milk to ferment, spoil, or grow dangerous pathogens. This is true for both raw and pasteurized milk, which is an effective growth medium for pathogens and spoilage microorganisms that can enter the milk after it is pasteurized.

One of the main nutrients in milk is the sugar lactose. Lactose is metabolized by bacteria through fermentation into two simple sugars and then into lactic acid and other products. Mostly beneficial bacteria are responsible for this type of fermentation and are called lactic acid bacteria or LAB. LAB are truly wonderful microbes; the acid they produce creates an environment in which not many other bacteria, including pathogens and spoilage types, can survive easily. So while the LAB continue to increase in numbers, the other types are limited, and eventually the LAB outnumber and outcompete them. In addition, by breaking down and converting lactose into lactic acid, these helpful bacteria often can help people who are lactose intolerant consume and utilize fermented milk products. As I mentioned earlier in the book, for most of our history humans have been eating their dairy products fermented—long before fresh milk was a regular part of the diet. Indeed, milk would have fermented relatively quickly and without any work on the part of our prehistoric relatives, given primitive storage conditions and containers.

Grouping by Gram, Staining that Is

You might have heard the terms "gram positive" and "gram negative" used to describe certain bacteria. This term has nothing to do with metric weight but is named after its inventor, Hans Gram, who in Berlin in 1884 developed a method for categorizing bacteria based on the way they react to the application and attempted removal of a dye. This method reveals the composition of the cell wall—one of the most distinguishing characteristics of bacteria. Unfortunately for us lay folk, categorization as gram positive or gram negative doesn't translate as "good" or "bad" bacteria; it's not that simple. But you will hear the term, both in this book and in most other discussions of the topic, so it's good to be aware of this common means of describing and labeling bacteria.

Understanding the Growth Rate of Bacteria

Once you understand the conditions under which bacteria will grow in milk, it becomes important to grasp that they do not grow at a steady, even rate of increase. Like the pH scale, their growth increases exponentially. In a nutshell, once they begin to grow, each generation will show a huge increase in numbers.

By understanding the growth patterns of bacteria and the conditions that must exist for optimal growth, you can begin to appreciate the importance of shelf life and storage conditions for fluid milk.

When bacteria are first exposed to the right conditions for growth, they go through a slow, almost zero growth period called the lag phase. During this time they are basically processing nutrients and preparing to grow. After the lag phase the rapid, exponential growth (also called logarithmic or log phase) occurs. But at some point the growth becomes self-limiting, since the bacteria will have consumed all the available nutrients. As they reach this point, the growth slows; this is called the stationary phase. After that the bacteria begin to die off in what is called the mortality or death phase.

By understanding the growth patterns of bacteria and the conditions that must exist for optimal growth, you can begin to appreciate the importance of shelf life and storage conditions for fluid milk. Even under optimal storage conditions, psychrotrophic bacteria may reproduce to a limited extent. If milk warms up, even just a bit, say from sitting out on the counter and is then returned to the refrigerator, you might have provided enough time for the lag phase of growth to be

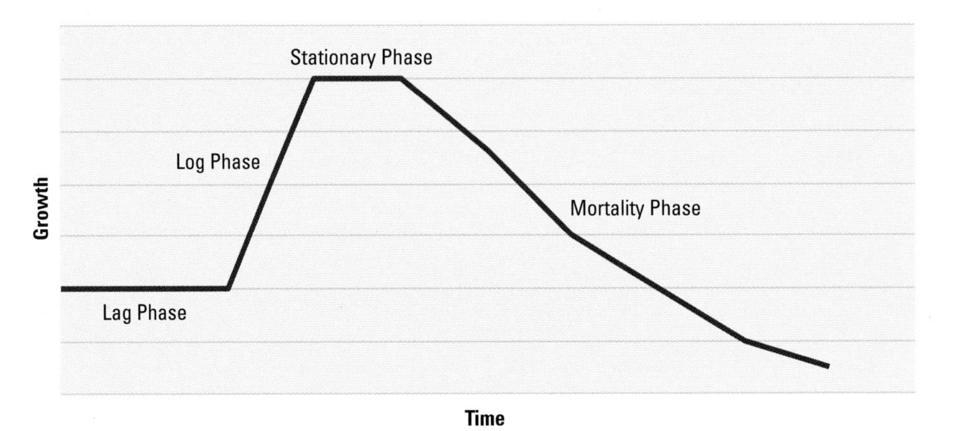

Typical grwoth phase pattern of bacteria

completed. If the milk is then once again allowed to sit out and warm, the bacteria may enter the exponential phase and grow to unwanted or dangerous numbers.

Bacterial By-Products

Now that you know more about the life of bacteria, let's talk about some other things that can be produced or created by microbes in milk. We have already talked about the lactic acid produced by many beneficial bacteria, especially during the purposeful fermentation of milk into things such as kefir, yogurt, and cheeses, but what other things can bacteria produce, and do these by-products matter for raw-milk quality?

Bacteria contain many different enzymes (proteins that speed up chemical reactions). As the bacteria live, reproduce, and die, they contribute these enzymes to the substance that has nourished them; in this case, milk. Since enzymes cause reactions—both destructive and constructive—they will alter milk from its original state. In some cases, such as in cheesemaking, this is necessary for the breakdown of proteins and fats into tasty substances and wonderful aromas. For fresh milk, though, it might mean damage to these same substances in a way that causes flavor and aroma defects (such as the "goaty" characteristics sometimes found in goat's milk and goat's milk products). On a positive note most lactic acid bacteria are known to produce lactase, which is the enzyme responsible for breaking milk sugar, lactose, into two simple sugars that all humans can digest. It is believed that bacterial lactase in raw milk is a possible reason some people who believe they have a certain level of lactose intolerance are able to digest raw milk.

At the other end of the spectrum of bacteria by-products are **toxins.** Toxins are the means by which some pathogenic bacteria cause disease and illness. Bacterial toxins are among the most powerful poisons known to man. Unfortunately for those of us who love raw milk, a couple of the bacteria that produce toxins can be found in milk (we'll talk about them later in this chapter). Keep in mind that it is extremely rare, but not impossible, for these pathogenic bacteria actually to exist inside the udder of the cow, goat, or sheep; but contamination is far more likely to occur during or after milk collection, rather than come from within the udder.

Gas is often the byproduct of bacterial fermentation. A common gas produced by bacteria, as well as other microbes such as yeasts, is carbon dioxide. Although some carbon dioxide production by purposely used bacteria cultures in cheeses is desirable, some is not. Coliforms are a large group of carbon-dioxide-gas-producing bacteria that are present at the farm in soils, bedding, and manure. Although most coliforms are not harmful to humans—indeed, many provide a valuable service to our digestive system—you'll learn more about some variations that can be deadly. In addition, the presence of coliforms is considered to be indicative of the possible presence of pathogenic bacteria. Microbes whose occurrence correlates with harmful types of bacteria are called indicator organisms. The presence of coliforms can be determined by the rapid production of carbon dioxide. You will learn in chapter 10 about a simple test that can be done on milk to determine the presence of coliforms.

RISK FACTORS ASSOCIATED WITH CONTRACTING A FOOD-BORNE ILLNESS

Most of us consider it common knowledge that when traveling to a less developed part of the world, we shouldn't drink untreated water or eat uncooked produce (as it has likely been washed in the untreated water). Not only do some of these vacation locales not have the same sanitation practices that we are accustomed to, such as purified drinking and vegetable-washing water, but we have not been exposed to many of the microbes to which the residents may have built up resistance. Here are some risk factors associated with susceptibility to illness from eating any food containing even small numbers of pathogenic bacteria:

- **Age:** The young and the elderly are more vulnerable. The young are vulnerable in part because of their lack of exposure and still-developing immune systems; and the elderly are vulnerable in part because of the increasing stress on their immune systems and their general vigor caused by the natural stress of getting old.
- **Pregnancy:** No matter how healthy a pregnant woman is, her body is working hard on a separate project—growing a baby—putting her body in a constant state of stress.
- **Recent illness:** If your body has recently pushed its immune system to the limits, it may not be ready to handle a new challenge.
- **Chronic illness or condition:** Increasing numbers of folks have an ongoing immune system challenge such as fibromyalgia, multiple sclerosis, arthritis, Graves' disease, or HIV/AIDS. Please remember that these risk factors vary greatly in their importance depending upon the person's longterm health and immune system function. Unfortunately there is not "rubber stamp" approach to assessing risk.

Microbial Maladies

The microbial world is filled with organisms that are beneficial to human health and also with a few pathogens that can cause life-threatening illnesses. Now that you have a good overview of microbial life, you probably can imagine that pathogenic microbes vary greatly in their ability to adapt to their environment, the number of microbes needed to cause illness, and the severity of symptoms they might cause in a person who is exposed to them. Always keep in mind that the presence of a pathogen does not guarantee illness!

In general the human population is believed to be more susceptible currently to illness from microbes than in recent history—largely because of a growing population of elderly, very young, and people with varying states of nutritional health and immune system compromises such as asthma and allergies, as well as more serious autoimmune diseases. In addition, the beneficial, immune-strengthening exposure to small doses of microbes throughout our lives, but especially as young children, has decreased tremendously. For the sake of your awareness and knowledge, let's briefly touch upon the current most concerning pathogens that might be found in milk. For an in-depth official summary of the most common culprits of food-borne illness, you can see the FDA's publication *The Bad Bug Book*, which is available online. Although it is important to know what health officials believe to be of concern, it is equally critical that the reality of the risks—the actual number of occurrences related to raw milk—are not necessarily in proportion to the public campaign.

Earlier in the chapter I talked about a common type of bacteria called *coliforms*, a usually harmless and actually even helpful group of bacteria present in the environment and in the lower digestive tract of humans and animals. However, there is also an increasing number of very toxic strains of one specific type of coliform called *E. coli*. The toxic strains can cause serious, even life-threatening conditions and possible lifelong damage for those who survive an infection. This category of *E. coli* produces a toxin called shiga. Of the shiga toxin (Stx) *E. coli* (STEC), the most infamous strain is 0157:H7 (said Oh-one-five-seven-H-seven), but other serotypes (that is the fancy word for "different strains") have been known to produce the same toxin, too. In fact, as of this writing there were six other serotypes (dubbed "The Big 6") known to have produced shiga toxin.

All shiga toxin-producing *E. coli* also are referred to as enterohemorrhagic or EHEC because they cause bleeding in the intestines (among other things). Although most pathogenic *E. coli* can cause diarrhea and digestive problems that are self-limiting, meaning they run their course without the need for medical intervention, those who are unfortunate enough to become infected with 0157:H7 may not be so lucky.

One of the most disconcerting things about STEC is the extremely small dose that health officials believe is required to cause illness in a vulnerable person. It is thought that as few as ten to one hundred cells may be enough to cause serious illness. This is part of the problem that producers face when testing milk for *E. coli* and specifically 0157:H7 (you will learn more about testing in chapter 10). A very small sample of milk (1 milliliter, roughly less than ¼ teaspoon) is tested. Picture 100 gallons of milk with even a thousand cells of *E. coli* 0157:H7 floating around. Guess how many 1 milliliter samples you could draw out of that 100-gallon tank? There are 3,785,541—yup, just under 4 million. This makes it very difficult to conclude assuredly that milk is free from the bacteria. In addition, one cell can quickly multiply if left to sit at room temperature or even in a refrigerator. So a glass of milk or a bowl of cereal could fairly quickly go from low risk to high for a susceptible person, while the remaining refrigerated milk remains unchanged. The importance of an unbroken chain of chilling cannot be over emphasized. Routine and proper testing for coliforms, and specifically *E. coli*, while not conclusive are still an important part of informing the producer of the presence of microorganisms that might indicate the presence of the pathogenic strains.

While *Listeria monocytogenes*, often simply called listeria, sickens fewer people than other food-borne pathogens (and is rarely linked to fluid raw milk related incidents), it causes a higher death rate. As with many "bad" bacteria, listeria infection, called listeriosis, is more likely to occur in those whose immune systems are not functioning adequately or who are under stress. Pregnant women seem the most vulnerable, with the infection potentially resulting in miscarriage or stillborn births. The tragedy of this type of outcome helps makes listeria a high-

PROBIOTIC PROTECTION

If there truly is a miracle food worthy of our attention, it might well be yogurt—and more precisely the probiotic bacteria that yogurt often contains. With names such as *Lactobacillus bulgaricus* and *acidophilus*, *Bifidobacterium longum* ssp *infantis*, and *Streptococcus thermophilus*, reading the label on a yogurt container can be akin to taking a science quiz. But these multisyllabic microbes are getting lots of attention for their possible and proven benefits to human health and well-being.

Not all yogurts contain the same bacteria, and not every yogurt bacteria culture is considered probiotic (meaning they survive the trip through the first part of our digestive system to reside in and provide benefits at the next, lower levels of our food processing centers). In repeated scientific studies oral probiotics, often fed in the form of yogurt, show significant abilities to limit and even prevent infection by food-borne pathogens. Not only do they provide this protection directly in the gut, but they also stimulate and activate the host's immune system.

profile pathogen. Listeria are common microbes found in soil and are present on farms of all kinds. They are tolerant of the cold, preferring to hang out in cool, damp places, such as animal water troughs and refrigeration unit compressors. Listeria are also quite tolerant of salt.

They have a reputation for surviving for long periods in facilities, despite the efforts of manufacturers to remove them. It is wise for any dairy producer to be at all times mindful of listeria's likely presence in the dairy and to take serious steps to prevent it from entering processing areas and product. The number of cells a person must ingest to make her sick is unknown, but it is believed to be very small, maybe a thousand for a person who is especially susceptible. Healthy adult nonpregnant individuals are believed to be capable of resisting infection in most cases.

Campylobacter jejuni, sometimes called "campy," is a normal resident in the gut of livestock used for food, including cows, goats, and sheep, but is most common in poultry. *C. jejuni* is estimated to be the third leading cause of all types of food-borne illness, according to the Centers for Disease Control and Prevention (CDC) and luckily is usually self-limiting and rarely fatal. It is, however, sometimes linked to the later development of Guillain-Barré syndrome (a crippling, sometimes life-threatening autoimmune disease). It is generally believed that it takes the ingestion of about ten thousand *Campylobacter* cells to lead to an infection, but some tests show that as few as five hundred may lead to disease. Again, the vulnerability of the person is an important factor. Interestingly enough, while the especially young are more susceptible, those aged fifteen to twenty-nine also seem more prone to developing symptoms after exposure.

Staphlococcus aureus is a common environmental pathogen, residing in dairy barns, soils, bedding, and so on. It can cause mastitis, an udder infection, in milking animals, and some methicillin-resistant strains are untreatable by antibiotics (it was the first major pathogen to become resistant to antibiotics, but an increasing number have done the same). Although staph is an uncommon pathogen in unprocessed milk and a low cause of death, the symptoms from a food-borne infection are unpleasant, thanks to the toxins created by this bacterium.

Salmonella is another common farm bacterium but is not a common problem in milk. As with other pathogens, however, it is easy to cross-contaminate raw milk from a source containing salmonella. Poultry is a common culprit—live birds, eggs, and meat (a big reason you don't see chicken on the sushi menu or raw eggs in an Orange Julius anymore). Handling poultry, even feeding your chickens, should not be followed by processing milk. The illness can last a few days but is self-limiting in otherwise healthy people.

Brucella, the cause of brucellosis, also called Bang's disease or undulant fever, used to be one of the leading health problems for cattle and for the consumers of their unpasteurized milk. It can occur in goats, but it is far less common in cattle. Although many states in the country are classified as "brucellosis free," the disease and the microbe still exist both in livestock and in wild herds of elk and bison. Fortunately, it is relatively easy to test animals to verify a "brucellosis free" herd. Dairy cows should be vaccinated while young. A brucella infection begins as a blood infection and causes intermittent fevers and sweating. The disease can become chronic.

While tuberculosis, or TB, is quite rare today, it is still considered enough of a risk that most herds, even in "tuberculosis free" states, that are producing raw milk for human consumption must be tested or certified free of the disease. As with brucellosis, TB was a leading human health concern up to the early to mid-1900s in the United States; milk was contaminated through human contact with the disease or infected cattle. As recently as the last decade, cases linked to dairy products have occurred, primarily through the eating of milk products produced in other countries where the disease is more common. Unfortunately, there also have been recent cases in dairy cows in California, Arizona, and Michigan.

Q fever, caused by the bacterium *Coxiella burnetii,* is a growing concern for raw-milk-borne illness. The bacteria can be passed in a number of fluids from infected animals to humans through direct or indirect contact (the bacteria can become airborne). Although most people do not become ill, some can become acutely or chronically ill. At this time I am aware of only one state, Washington, requiring testing of animals that provide raw milk for human consumption. Although transmission by milk is considered rare, this disease is of great concern for herd managers because of the easy ways it can spread throughout a herd and even to neighboring animals and humans through direct or airborne contact with birth fluids. In fact, it is *C. burnetii* that caused the then United States Public Health Service to increase the vat pasteurization temperature minimum from 143°F (62°C) to 145°F (63°C) after it was demonstrated in the late 1950s that *C. burnetii* could otherwise survive. Washington State has developed best practices that are meant to control the spread of Q fever (see the bibliography).

When learning about milk-borne pathogens and diseases that can be transmitted from animals to humans (called zoonoses), I recommend trying to step back and look at the big picture. It is easy to become so focused on risks that you lose perspective. At the same time one must never lose respect for the reality of what these illnesses and diseases have done—and can still do to humans today. It is also critical to look at things in the light of developing issues and changing paradigms. As I was finishing this manuscript, the Centers for Disease Control and Prevention issued a report called *Antibiotic Resistance Threats in the United States, 2013*, in which they stated that at least 2 million people a year become infected with bacteria that is resistant to all known antibiotics, and at least 23,000 of these people die from that infection.

Compare this to the 2011 CDC estimates of total food-borne illnesses (this includes simple "food poisoning"): 48 million people are sickened, 128,000 are hospitalized, and 3,000 die of food-borne diseases. That is less than seven times the number of fatalities from antibiotic resistant bacteria, and the vast majority of these illnesses do not require a visit to the doctor. During that same year there were approximately sixty cases of illness related to raw milk and raw-milk products—and no deaths. With the CDC warning of the approach of a postantibiotic era and studies pointing toward naturally boosting our immune systems through exposure to microbes, I believe we are ready for a new germ theory.

· 5 ·

Udder Understanding

Now that you have a solid understanding of the basics of microbiology, let's take a step back and look at how milk is made inside the animal and how it is harvested by both the young of its own kind and humans. There's an old saying that I am fond of repeating as a way to emphasize a key factor about milk safety and quality. The axiom states, "Milk was never meant to see the light of day." In other words, milk was designed by nature to be consumed immediately and to enter the stomach, where it is quickly acidified and digestion commences. Once our ancestors decided that milk was something to be collected, stored, and used later, there were many opportunities for the quality to be compromised.

Without acknowledging these potential problems and knowing what can be done to limit them, you cannot be a responsible drinker or harvester of milk. Let's begin by taking a look at the amazing production of milk—from the inside out. Then we'll apply the basic microbiology from chapter 4 to help you understand the challenges facing the collection of high-quality milk and the steps that must occur to give it the best chance at remaining a pure, superior food.

An Inside Look at Making Milk

Dairy animals produce and store milk in a body part called the udder. Old-fashioned terms for the udder include milk sack and bag (a well-known udder ointment in an iconic green tin is called Bag Balm). Milk is removed from the udder by the baby or the human through the animal's teat. The entire milk-producing anatomy of the animal is called the mammary system.

A cow has four teats and four separate milk-producing portions of her udder called quarters. A doe (milk-producing female goat) and a ewe (milk-producing female sheep) have two teats and, accordingly, two udder compartments called halves. Each teat is supplied milk from its corresponding quarter or half. This is important to know, since an animal can have an udder infection in only one portion of the udder. Each quarter or half must almost be thought of as producing milk from a separate milk source.

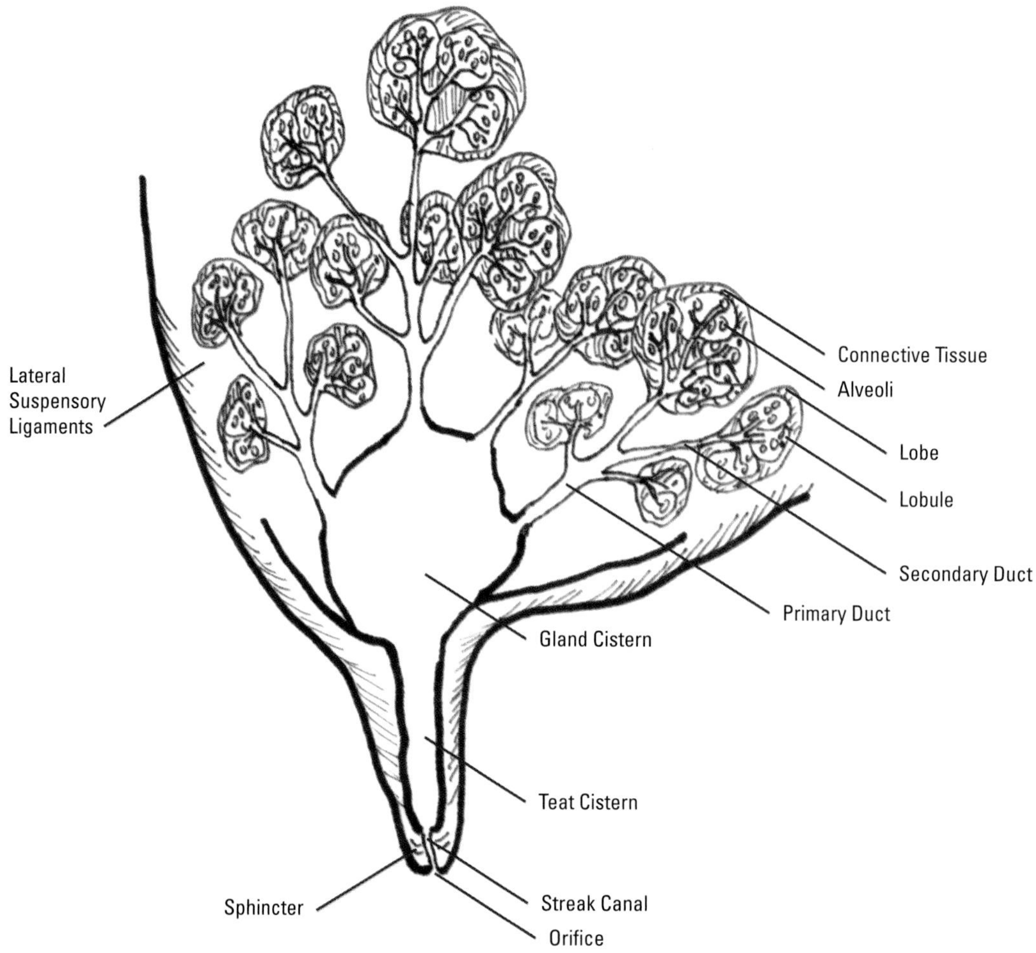

A cross-sectional illustration of an udder.

The end of the teat has a little opening called the orifice or streak canal. The orifice has a circular muscle around it called the sphincter whose job it is to keep the milk in and bacteria out. After milking, it takes about ten minutes for the orifice to tighten back up and do its job; it's during this period of time that the udder is at greater risk for harmful bacteria entering the teat and potentially moving up into the udder. Ideally the animal should stand in a clean area during this vulnerable period, but to help ensure protection many farmers apply a protective, sanitizing, moisturizing teat dip right after milking that helps provide temporary protection. This is especially helpful in wet weather or conditions in which the animal might splash manure back onto her udder. To help prevent the animal from lying down in damp areas, some farmers encourage their milk animals to go from the milking parlor to a feed area where they can stand and eat while their teat orifices tighten back up.

As an animal ages and has been milked for several seasons, the orifice can grow more open and vulnerable to infection. Although we dairy farmers enjoy hand-milking animals with more open orifices, since it is easier to get the milk out, over time these animals might be more prone to mastitis. The orifice can be damaged

MYSTERIOUS MILK: WHEN KIDS, BUCKS, AND NONPREGNANT DOES MAKE MILK

Goats are funny animals in many ways (part of the reason they are so endearing to many of us). One of their most unusual tendancies is the production of milk when you least expect it. This type of behavior is rare in cows and sheep, but in caprines, udders of various sizes might be seen on a newborn baby, a doe that has never had kids, or even a full grown, smelly, masculine buck. These conditions go by various names: "Witch's milk" is the term used when a baby has a small udder; "precocious udder" describes the development of all or part of the mammary gland on a young or aged doe that is not pregnant or has not given birth; and the strangest of all is gynecomastia, a condition in which a fertile buck forms an udder with milk. Usually the appearance of such an aberration can simply be ignored, but in some cultures the milk of a vigorous breeding buck is seen as an aphrodisiac.

by vacuum pressure that is too high, improper attachment of milking equipment, inadequate letdown response (more on that later in this chapter), and late takeoff of milking equipment.

Inside the teat is an open area called the teat cistern where milk pools between sucks by the baby or squirts by the farmer. The teat cistern is supplied milk by a larger pooling area just above the teat called the gland cistern (remember the udder is a gland—as all glands do, it secretes a substance). When milking an animal, you can determine when she is empty by feeling the area of the gland cistern. It will change from feeling full and squishy to empty and saggy.

The gland cistern is supplied by multiple little tubes called milk ducts that lead from the actual tissue that produces milk. The ducts continue to branch and grow smaller as they near the actual milk-production tissue. The milk is secreted in this tissue by individual cells lining small cavities called alveoli. The milk-secreting cells are constantly supplied by blood vessels from which they draw their constituents. Try to remember this close connection of the blood to the milk, as we will be talking about things later that make this relationship important for the quality of the milk. Cows, goats, and sheep with large cisternal areas tend to be able to store more milk in their udders, believed to make them good candidates for once-a-day milking—a common practice in several countries. I'll talk more about that option later in this chapter.

An important difference between the milk of cows, sheep, and goats has to do with the manner in which these different species make milk. All glands (one of the defining features of being a mammal) secrete their products one of three ways. Two of these are pertinent to milk secretion—apocrine and merocrine. Glands that secrete via the apocrine system also shed parts of the cell wall lining. Goats, sheep, and humans secrete milk via the apocrine system, while cow's milk is shed via the merocrine system, which keeps the secretory cell intact. It is for this reason that goats and sheep are allowed to have a higher somatic cell count (SCC), a test in which milk is analyzed for the presence of somatic—literally "originating from the body"—cells. This test counts white blood cells to determine if an infection is present in the udder, but it also counts the other body cells excreted during milking.

SMALL DAIRY PROFILE: HAWTHORNE VALLEY FARM, NEW YORK. CERTIFIED RAW-MILK COW DAIRY.

Dairy plant manager Peter Kindel describes biodynamic Hawthorne Valley Farm as "a raw milk dairy poster child," and after talking to Peter about the sixty-cow dairy with its Brown Swiss, Jersey, and Red Devon cows, I have to agree. Peter, a well-informed, well-traveled, and well-educated cheesemaker, takes a balanced, proactive stance regarding the production of high-quality raw milk for both cheesemaking and fluid consumption. The farm, located on 400 acres in Columbia County, New York, has been selling raw milk directly to customers from the on-farm store since 1989. Currently about 45 half gallons leave the farm every day, thanks to dedicated and eager farm patrons.

The Hawthorne Valley story is worthy of its own book. The farm was purchased in 1972 by a group of like-minded educators, farmers, and food artisans. In addition to raw milk and cheese, the farm produces fruits and vegetables, sauerkraut, and artisan breads. They also fulfill their educational mission by hosting multiple training, farm experience, and learning opportunities. While it seems as if 400 acres of farmland could easily support more cows than Hawthorne's herd of sixty, the biodynamic and permaculture approach to managing the land puts that number as just right. Although the farm brings outside milk in for the production of some of their other products such as cheese and yogurt, as a key part of their food safety plan, they use only their own milk for fluid sales.

Although the state tests the milk regularly, Hawthorne Valley sets the perfect example, in my opinion, of a testing protocol that every raw-milk producer should follow. On the morning that milk is to be bottled, the previous evening's milk is sent into production for other products. The bulk milk tank is then completely cleaned and sanitized, and then the cows are milked. After the milk chills, samples are taken, and lab tests for total bacteria, total coliforms, *E. coli*, aeromonas, and salmonella are begun. Once a month Kindel hands a handful of InSite listeria swabs to his team and challenges them to find listeria anywhere in the plant. They can't, thanks to the facility's cleaning procedures (more on InSite swabs in chapter 10).

Once test samples have been pulled, milk is bottled and moved to a holding refrigerator pending the on-site lab results—which take twenty-four hours using the ECA easy gel system. Antibiotic residue testing is done in the farm's certified testing lab. (Although the bacteriological testing is not certified, they regularly train other farms to perform such tests.) If the lab sample comes back less than stellar, the milk is rerouted to a pasteurized product and not sold raw. The state tests for such things as campylobacter, which has never come back positive, and the farm has never had to recall any milk. Not bad for a twenty-plus year career!

Because of the great care and attention taken, this farm is a beautiful example of how milk can be produced and provided in the manner it and consumers deserve. Whether you are milking one cow or a hundred goats, this farm's practices are definitely worthy of emulating. For more on Hawthorne Valley Farm, visit *http://hawthornevalleyfarm.org/*.

Most references say that milk in an udder without any infection (we'll talk more about this later) is sterile—meaning free of bacteria. Just to give you another body part as a frame of reference, urine in our bladders is also sterile (unless an infection is present). Research in humans suggests that in some cases beneficial gut bacteria from the mother are indeed transported into the breast milk through the mother's bloodstream; whether this is true in other animals is not yet known, but it is certainly conceivable. Either way, it is important to know that the udder is constructed to try to keep *external* contaminating bacteria out.

Feeding Their Young

Mother animals, except under unusual circumstances (see earlier sidebar **"Mysterious Milk"**) will produce milk only in anticipation of feeding a baby, so they must become pregnant. During the last month or two of pregnancy, the fetus sends hormonal signals to the expectant mother, and her udder begins developing milk-producing tissue. Between pregnancies the moms are given a rest period, usually no less than two months long, when they do not give milk. During this time they are called *dry.* During the dry period the milk-producing tissue in the udder is reabsorbed and the udder shrinks.

As birthing approaches, the udder produces a first milk called colostrum, a thick, yellowish to amber milk-like liquid filled with dense nutrients, high in caloric value, and, most importantly, chock-full of antibodies to protect the new baby from diseases and infections it will suddenly be faced with after leaving the safe and sterile environment of the womb. If the mother is unhealthy or otherwise unfit, she won't make high-quality colostrum and the babies will be more susceptible to problems, including sudden death at a few days of age.

When the baby nurses the mother, it also ingests a healthy dose of bacteria from the outside of the mom's teat, as well as some that have migrated into the teat canal. This bacteria (which many raw-milk drinkers hope to ingest as well) helps the baby begin establishing its own natural, intestinal flora and exposes the young animal to organisms that it must learn to build up resistance to as it grows—think of this phenomenon as oral vaccinations. (There is an old saying that human

In natural settings, young animals often nurse their mothers far longer than in most commercial operations.

children should "eat six pounds of dirt by age six" to be protected from all sorts of future ailments.) Colostrum production lessens and becomes more diluted by milk at a steady rate over the first day or so after birth, and it is mostly gone by day two and usually completely by day four.

To varying degrees the flow of milk is controlled by a response to the baby's suckling called letdown. The letdown response occurs because of the release of a hormone called oxytocin from the momma animal's pituitary gland. In some species the mother may produce only oxytocin and let down her milk when her baby is present and nursing; animals that do this include camels, whose milk is used for human consumption in mostly nomadic cultures, and water buffalo, whose milk is popular for the production of fresh mozzarella (*mozzarella di bufala*). This brings many challenges to those choosing to milk these beasts commercially. Most dairy cows, goats, and sheep will begin to let down their milk fairly readily from the simple stimulation of the farmer washing the udder and prepping the teats for milking. Some farmers, myself included, also massage the mom's udder a bit during milking to sustain the letdown and encourage the udder tissue to release all its valuable liquid.

When a baby suckles, it covers the teat with its mouth, then creates both suction (negative pressure) and a massaging action with its tongue. This combination pulls the milk from the teat and cistern; then the baby releases the suction, which allows blood circulation to flow in the teat tissue and lets the teat refill with milk. When farmers milk by hand, they use a manual squeezing and releasing to force the milk from the teat. There are many techniques for doing this. Most are quite gentle to the animal and can be performed in a very clean fashion. Modern milking machines simulate the negative pressure and massaging action of the baby and can also be quite gentle and clean. I'll cover the ins and outs of proper use of milking machines in chapter 8 and hand-milking techniques later in this chapter.

Mastitis

Mastitis, an infection of the mammary gland, is a costly reality for the dairy industry. Its severity ranges from mild, subclinical cases to severe infections that can cause all or part of an udder to become gangrenous and either slough away or cause the animal's death. In mild cases milk production is decreased and milk quality compromised, often without the producer even being aware that a subclinical case exists. Between loss of production, either temporary or permanent; the loss of animals; and the cost of treatment, every dairy farmer is concerned about what causes mastitis and what can be done to reduce its occurrence. Entire books are dedicated to this topic, as well as an organization, the National Mastitis Council. So it is a lengthy topic that I will only touch upon but will also address throughout this book whenever possible—and pertinent.

In the last chapter you learned that the udder can be infected with some of the same pathogens that can infect humans. Earlier in this chapter I talked about some of the natural differences in the number of somatic cells that occur in milk. Later in the book I will be teaching you more about testing your milk on the farm for somatic cells and how to interpret these results.

It is my opinion that mastitis concerns can be reduced greatly through preventive measures that boil down to good farm practices that you should implement anyway for healthy animals and quality milk. Through a clean environment, minimizing stress on the animal, healthy diet, and proper milking procedures—including equipment maintenance and cleaning and regular monitoring of somatic cell counts (SCC)—udder infections usually can be nipped in the bud while they are still "under the radar," as subclinical mastitis. When addressed at this level, organic approaches to treatment have a great chance of working. Here at our farm, after eleven years of milking (year-round for the most part), we have yet to have a single case of visible, acute mastitis. That is not to say that a severe case is not around the corner, but it is a pretty good record. I attribute this mainly to herd health and monthly monitoring of SCC followed by immediate measures taken when a count rises above normal.

STEPS FOR REDUCING THE LIKELIHOOD OF MASTITIS

1. Observe the first squirts of milk (foremilk) from each half or quarter of the udder at each milking for abnormalities such as blood, clumps, or stringiness.
2. Observe the udder for reduced production or heat in any quarter or half.
3. Monitor somatic cell counts from each animal on a regular basis
4. Monitor milking equipment for fit and optimal operation—including how it is used by those doing the milking.
5. Increase herd health and conditions for optimal udder health.

I encourage every dairy farmer to not believe that mastitis and the regular, therapeutic use of antibiotics (especially at dry-off time) are a normal part of having dairy animals. Indeed, the effectiveness of therapeutic levels of antibiotics, such as used in "dry treatment" are diminishing with more antibiotic resistance becoming common. This does not mean that antibiotics should never be deployed; they absolutely have their role to play in the prevention of suffering. But be aware that antibiotic-resistant staphylococcus can infect the animal's udder. In these cases, there is no treatment other than culling the animal from the herd. Hygienic udder preparation and milk collection practices will help limit the likelihood of spreading microorganisms between animals. Animals with chronically high SCCs should have milk cultured by a laboratory or on farm testing to determine whether culling is necessary. On farm culturing (OFC) can be done using so called "quad plates" designed to test a single milk sample for multiple mastitic organisms. See the resource section for more information.

Harvesting Milk for Humans

Earlier I mentioned that milk was never meant to see the light of day. But since milk is a delicious and nutritious food staple, there are many strategies that can reduce the likelihood that milk will lose quality as it is collected and stored. These

MILKING FREQUENCY—ONE, TWO, OR THREE TIMES PER DAY?

All mammary glands respond to signals from the body that tell them how much milk to produce. One of these signals is generated by the frequency of nursing and the amount of milk removed. The more milk drained from the udder, the more milk the mom will try to produce. Of course, there are limits to this production, and proper nutrition and other factors all come into play as well. In the United States it is most common on all sizes of farms to milk twice a day. Some larger farms rotate the milking stock through three times a day. In other parts of the world, though, once-a-day milking is becoming more common as a way to control labor and other costs and increase efficiency.

When compared to twice-a-day milking, most data suggests that daily milk volume will increase by about 10 percent when the frequency increases to three times in a twenty-four-hour period and decrease by about 20 percent when the frequency decreases to once a day. These are generalized numbers, however, since the animal's own anatomy must be able to accommodate the changes in volume.

Concerns over udder health when decreasing milking frequency are valid, but research indicates that if udder health is good and somatic cell counts are low, the once-a-day milking does not increase the occurrence of mastitis. If, however, animals have elevated SCC and decreased udder health, then decreasing milking frequency can increase the occurrence of mastitis. Given that one of the most effective treatments for mastitis is increasing the milking frequency—to keep the udder emptied of milk and bacteria—this is a logical conclusion.

My own experience with three times a day occurred when preparing our does for the National Dairy Goat show. We began milking every eight hours about a week or so before the show so they would to look their "milkiest" in the show ring. By the time of the show, their production had noticeably increased, filling their udders to the maximum. Animals that nurse their young continually or even part time, rather than being milked only by the farmer, also will show increased production.

It may seem counterproductive to reduce the frequency to once a day, but the decrease in labor, grain, and chemicals for cleaning may be worth the lost milk production for some producers, commercial or home. We switched to once a day after losing our evening milking help (my youngest daughter grew up and moved out), and I don't think we will ever go back. Milk production has held at 75 to 80 percent of what it was, and the animals adjusted quickly—with even the heaviest producer becoming comfortable within a couple of days. We have had no decrease in udder health either. Although the practice still garners much criticism in the United States, because of existing paradigms of what is "normal," data and practical experience on good-size cow dairies in New Zealand and Australia and high production goat dairies in Europe support the adaptation of this approach as humane, sustainable, and financially equitable.

strategies and tactics also help reduce the chances of unwanted, harmful bacteria growing in the milk. When milk is consumed raw, it is important to honor its delicacy and its life-giving nature—which appeals not only to us but to uncountable numbers of microbes. So let's go over the factors and principles of collecting milk, in the context of the microbiology we covered in the previous chapter, to understand the reasoning behind each step of milk collection and storage.

Clean Animals

When milking animals come into the parlor, they should be relatively clean and dry—especially around their udders, flanks (part of their bodies just above the udder), and any part of their bodies that will be above or near the milking pail or milking machine parts (known as "the milking zone"). Remember that even if you cannot see it, minute particles of manure and dust from dirt and bedding will be floating in the parlor and landing in the pail or will be sucked into the milking machine system (which contrary to popular thought is not a closed system). If the animal is wet, either from being outside or from being washed down, the moisture will draw contaminants and possibly drip into the pail or be sucked into the milking machine.

To maintain animal cleanliness many dairy farmers will keep the udder and the surrounding anatomy clipped, shaved, or, on some big dairies, singed free of excess udder hair using a special propane torch. Animals also can be brushed or washed before entering the parlor. If washed, great care must be taken not to let any of this wash water become a part of the milk collection. Of course, animals will come into the parlor cleaner if they have dry, clean places to rest and sleep. Goats, unlike sheep and cows, are particularly easy to keep clean, as their manure is dry and they do not like rain or mud and will avoid wet situations whenever possible.

Clean Teats

The skin of the teat, including inside the teat, is the natural habitat for some beneficial and harmless bacteria—many of which can be helpful in the production of cultured and fermented raw-milk products. But it is also the home to pathogens and spoilage bacteria from the environment. Therefore, teats must be thoroughly cleaned before milk is collected. The following steps apply when animals come into the parlor relatively clean, an issue addressed earlier. Please know that these steps are sometimes slightly reordered, with udder cleaning taking place before stripping.

CLEANING THE UDDER AND MILK LETDOWN

As we covered earlier in the chapter, when the baby animal nurses its mom, the presence of the baby along with the stimulation of the teat and udder (the babies often butt and nudge the udder with their heads) sends a signal to the mother's brain that causes the release of oxytocin. This hormone is useful for many things in the animal's life, but in this case it signals the udder to "let down" the milk from the milk ducts. On the dairy farm you can encourage this response by removing the first few squirts of milk (prestripping) and cleaning the udder. In cows this response is profound, with the udder going from partly full to fully engorged in a few moments. Dairy goats have a less dramatic response, the milk being released in a more sustained fashion. Dairy water buffalo will let their milk down much better if their calf is present, making management of these massive dairy animals more challenging.

For the sake of udder health, milking machine equipment needs to be attached to the udder soon after the letdown response is initiated. Part of the response causes the teat sphincter to relax and allow the milk to more readily exit the teat. If this response diminishes and the sphincter tightens, machine milking might increase the chances of physical trauma to the teat orifice, leading to increased somatic cell counts and vulnerability to mastitis.

TWO APPROACHES TO UDDER PREP

Method A—Hand or Machine Milking

1. Clean the entire udder with a hot, soapy but wrung-out towel.
2. Remove three squirts of foremilk. Observe for abnormalities such as blood, clumps, or stringiness.
3. Dip each teat with an approved teat dip—coating the teat thoroughly. Let it sit for thirty to sixty seconds, or as recommended by the manufacturer.
4. After thirty to sixty seconds, dry the teats with a single-use towel (paper or washable).
5. After milking, dip the teats again with an approved teat dip.

Method B—Machine Milking, Udders Clipped and Free from Debris

1. Dip each teat with an approved teat dip—coating the teat and milking zone. Let it sit for thirty to sixty seconds.
2. Remove three squirts of foremilk. Observe for abnormalities such as blood, clumps, or stringiness.
3. Dry each teat and zone with a single-use towel.
4. After milking, dip the teats again with an approved teat dip.

The first step in udder preparation is the removal of several squirts of milk from each teat—about 2 tablespoons (about 30 ml)—called the foremilk. This step also is called stripping and usually is done using a strip cup to catch the milk and observe it for abnormalities. The foremilk will contain a great number of microbes that will have moved into the teat in between each milking. Many farmers clean the udder first, then strip the teat, but most current advice recommends stripping first, then cleaning to limit the possibility of some of the solution remaining on the teat and getting into the milk. But if a predip follows the full udder wash and stripping, it is perfectly acceptable to strip after the udder wash.

After you strip the foremilk, you must clean the milking zone thoroughly. The milking zone is any place on the udder that will make contact with your hands or with the milking machine. If you are milking the animal by hand, you should clean the entire udder (this can take place before or after stripping, as noted earlier) with a soapy, hot water solution—making sure that the udder is not dripping wet afterwards. This is followed by an even more thorough cleaning or dipping of the teats and udder tissue just around the teats. You can use a chemical antibacterial such as iodine, chlorhexidine, chlorine, quaternary ammonia, hydrogen peroxide, or acid-based sanitizers in conjunction with a skin softener (to help prevent chapped teats) and surfactant to remove dirt, sanitize, and condition. Some farmers use a cloth or teat wipe to clean the teats with these solutions while others use them as a "predip." Predips are applied by cup or sprayer and allowed to have contact with the teat for at least thirty seconds to ensure maximum effectiveness.

Whether sprayed or dipped, the solution must make contact with the entire surface of the teat that will come into contact with the milking machine teat cup. For this reason coloring is often used in dips, so you can see if the coverage is complete. If you use a cloth to apply a solution and thoroughly hand-clean, you can shorten the contact time of the dip or udder wash because the mechanical action of washing will help reduce dirt and bacteria counts on the teats. On large farms it is usually faster to teat-dip animals than it is to hand-clean.

Once the cleaning or dipping is complete, use an individual, single-use dry towel to remove any remaining solution and moisture. These can be disposable paper towels or small clothes that can be laundered and reused. If the entire udder needs to be dried, the teat should be dried first when the towel is at its cleanest.

After milking, many dairy farmers apply a "postdip," often the same solution that was used to predip. This postdip is intended to protect the teat from contaminants immediately after milking. After ten to twenty minutes, the orifice will tighten back up and help prevent most contaminants from entering the udder. Some farmers encourage the animal to stand for that period of time, instead of postdipping, by providing hay or feed just outside the parlor. The combination of antibacterials used before and after milking can greatly reduce the incidence of mastitis and can lower somatic cell counts.

The standard practices of udder cleaning are based on the goal of removing dirt, debris, and as many bacteria as possible. Although I am not advising you to not follow this protocol, you should be aware that there is some thought and ongoing discussion that may one day take udder cleaning in a new direction. In chapter 10 I will cover a test called lacto-fermentation that is used to determine the presence of desirable lactic acid bacteria. The natural presence of these bacteria on the teat, naturally occurring in far greater numbers than spoilage or pathogenic microorganisms, might be better managed through the purposeful support and collection of such bacteria, rather than the attempted sterilization of the teat surface. The presence of these microbes is greatly dependent upon environment and the animal's adaption to her living conditions, as well as exposure to sanitizers. Here

TABLE 5.1. Some Commercially Available Udder Wash and Teat Dip Solutions*

Brand Name	Use	Active Sanitizing Ingredient	Notes
Spectrum	Udder wash	Chlorhexidine and quaternary ammonium	Chlorhexedine is antibacterial. Quaternary ammonium is also antibacterial but is also used as a surfactant.
Fight-Bac	Spray-on teat sanitizer	Chlorhexidine	Antibacterial teat spray
CHG	Pre- and postdip	Chlorhexedine and quaternary ammonium	See Spectrum, above
DermaSept, Derma Pro, Pro Guard	Postdip	Capric and Caprylic	This environmentally friendly product uses the same fatty acids that occur naturally in milk to provide antibacterial action.
Teat-Kote, Foam-N-Dip, and many others with varying concentration of iodine and emollients	Pre- and postdip	Iodine	Iodine is an old standby for udder sanitizing. Expect staining in plastic milk hoses. Be sure to dilute properly to limit iodine residues in milk, and dry teats thoroughly.
Oxy-Gard, Oxy-Pro	Pre- and postdip	Hydrogen peroxide, lactic acid	Environmentally friendly, in combination with lactic acid helps improve teat skin condition
Oxy Dip	Pre- and postdip	Hydrogen peroxide, sodium lactate (lactic acid)	Same ingredients as above, but not with FDA-approved label

** See appendix A for suppliers.*

WHAT ABOUT HOMEMADE UDDER WASHES AND DIPS?

If you are a commercial producer, it is wise to stick to products with labeling accepted by the FDA. That said, there are plenty of commercial products that have identical components but are not FDA accepted, such as Oxy Dip in Table 5.1. In the case of products like this, you'll need to decide for yourself whether you are comfortable using them.

During our years of milking, I have mixed many of our udder washes and sanitizers myself. I paid close attention to how these procedures and solutions affected the bacteria counts in the milk and the condition of the teats. It doesn't take much to cause teat irritation with an improper sanitizer concentration. Irritation can lead to infection, high SCC, and poor-quality milk—not to mention discomfort for the animals. Teats in poor condition will also not milk out as quickly and efficiently.

When mixing sanitizers, whether for udder or equipment cleaning, proper dilution is not simply a matter of measuring. The amount of water you mix with the chemical will have an effect on how well the antibacterial works. For this reason it is critical to use test strips designed to measure the type of sanitizer. In addition, you can't mix teat conditioners with sanitizers without knowledge of how the two will interact. For all of these reasons, I've determined that it's a better use of my time to purchase properly mixed and formulated solutions for anything containing a sanitizer. For simple washing of udders, though, you can use a gentle dish soap of any kind, combined with hot water, especially when followed by a properly mixed teat dip and single-use drying towel.

A common homemade recipe for udder wash calls for a bit of chlorine bleach and Dawn dish detergent. In theory this should be an effective wash and sanitizer, but the label on Dawn specifically warns about combining the two ingredients. Although Dawn does not contain ammonia, which most people know cannot be mixed with chlorine because a dangerous chemical reaction will occur, it does contain other chemical compounds (namely ethanol alcohol) that interact with chlorine and produce unhealthy and even explosive gases. You may not notice them, but they will be there. Remember that all detergents and soaps are not the same! You must **read and respect warning labels** and the impressive nature of chemicals to react and change.

at our farm we are experimenting with a different approach to udder cleaning that does not involve strong antimicrobials and sanitizers. The effectiveness of such an approach is being monitored with bacteria counts and lacto-fermentation tests, all of which you will read about in chapter 10. Although very little research is being conducted on such approaches, especially here in the United States, it holds great promise not only for collecting healthy microbes, but in possibly helping prevent udder infections through the support of natural skin bacteria.

Clean Parlor

The milking parlor or area is the first place where milk will make contact with the environment. It is also the first opportunity for contamination. Dust, yeasts and molds, bacteria, viruses, hairs, and manure *will* be in the air of the milking area. You must make every effort to limit the milk's exposure to contaminants. Your first line of defense was mentioned earlier—making sure that animals entering the parlor are as free from filth and excess hair as possible. The next step is cleaning the parlor thoroughly after every milking. During milking, be aware of anything occurring around the area that might stir up debris that will then drift into the parlor—from pen cleaning to windstorms. Milking in open pails can expose milk to debris, but milking machine systems also pull air into the lines when operating. Despite appearing to collect milk without contact with the outside environment, mechanical milking systems are not closed systems.

Clean Equipment

Although this may seem like an obvious goal, keeping milking equipment clean is an ongoing challenge for dairy producers. The nature of the equipment—with long hoses, tiny parts, and many nooks and crannies—along with the nature of milk—with fat, protein, and minerals being deposited on all surfaces—makes keeping this equipment clean and sanitary a challenging task. Plenty of hot water, time, physical scrubbing, and the right cleaning and sanitizing chemicals are a must, as well as the replacement of parts on a set schedule, depending on what they are made of. Long before they wear out, parts of a milking system will begin to harbor and grow bacterial colonies, leading to the contamination of each batch of milk that flows past. In chapter 9 I'll cover the exact steps needed for cleaning and maintaining dairy equipment.

Proper Milking Technique

Whether you milk by hand or by machine, proper milking technique is important for both the comfort of the animal and the long-term health of her udder. Either approach can be gentle or it can be damaging. Most of us first learn to milk by hand, and for many people, spending this time with their animals is one of the most enjoyable and rewarding parts of dairying. When milking more than one cow and a handful of small ruminants, however, milking time can

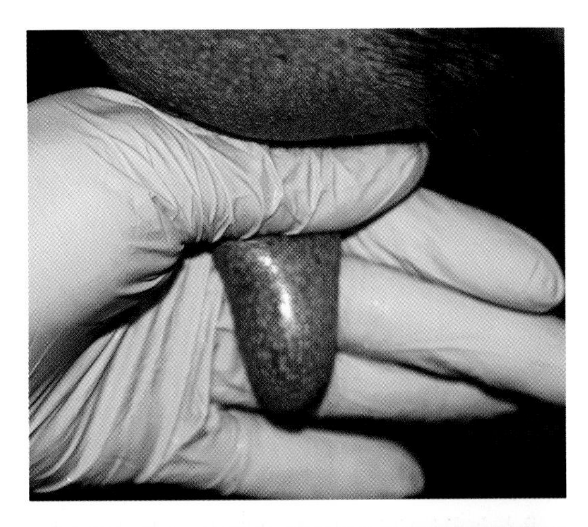

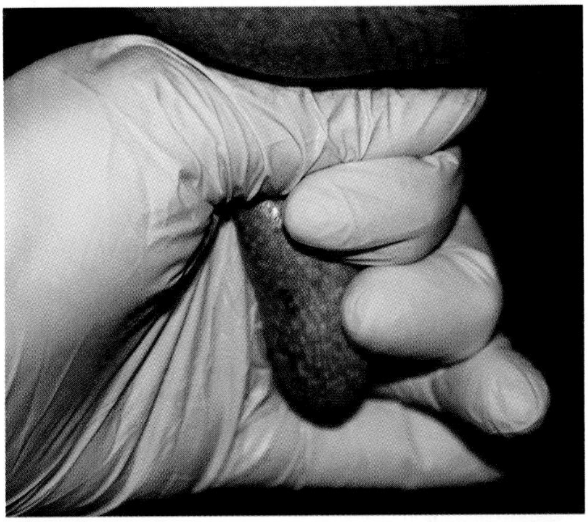

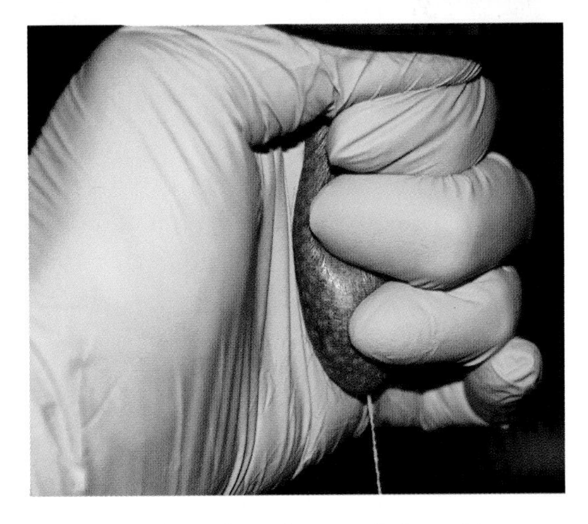

Hand-Milking, Step 1: Grasp the teat at the base of the udder, using your thumb to tightly trap milk in the teat. Step 2: Roll your index, then middle, then ring finger in to force the trapped milk out through the orifice. Do not pull down on the teat while doing this. Step 3: Release pressure, and allow teat to fill with milk. Then repeat all steps.

often pass from pleasurable to tiresome and inefficient. In addition, milking by machine offers some opportunities to prevent environmental contamination of the milk—if the milking equipment is properly cleaned and sanitized.

There are several techniques for milking by hand, some that might seem exotic to Westerners. If you learn a method that differs from the one I suggest here, that is fine; just keep in mind the factors that might make any method irritating for the animal, and for yourself. If you are new to hand milking, give yourself plenty of time to practice on a patient, well-trained animal. I find it usually takes even the best learner about three days to master the method, and a few days to weeks longer to develop enough speed to graduate to what I call the "cappuccino category," in which the milk pail is topped by a thick layer of dense foam.

Milking machine equipment—the cleaning and maintenance of which we'll cover in greater detail in the next part of the book—must fit the animal, be placed on the teat properly, have the right vacuum level, and be removed properly. Although mechanical milking machines used to have a reputation for causing udder damage, this is unfounded when they are used properly and monitored. In fact, milking machines more closely mimic the way that a calf, kid, or lamb removes the milk from its mother, and most milking animals seem to find the milking machine process quite comfortable.

COMMON MILKING MISTAKES

Hand-Milking

1. Grasping too high on the teat and damaging the udder tissue
2. Pulling downward while milking
3. Milking with wet hands (which spreads bacteria into the milk and teat)
4. Using a "stripping" technique, in which the thumb and forefinger are used to slide milk out of the teat, causing irritation and inflammation inside the teat canal

Machine Milking

1. Vacuum pressure too high or too low (causing the teat cup to slip up and down on the teat)
2. Improper rate or ratio of pulsation
3. Placement of inflation on teat too high so that it encompasses udder tissue
4. Slipping up and down of inflation on teat, causing vacuum fluctuations
5. Worn or improperly fitting inflations
6. Leaving equipment on too long after milk flow has ceased
7. Unstable milking system vacuum level (causing slipping and vacuum fluctuations)
8. Removal of inflation without turning off suction

Filtering and Chilling

Once you've collected the milk, you must filter and chill it to refrigeration temperatures, below 40°F (4°C) as rapidly as possible, generally less than two hours after milking. You might recall from the previous chapter that bacteria experience a no-growth, or lag, phase. It is during this phase that chilling should take place so that little to no bacterial growth can occur in the milk.

Note that, even if you use a milking machine, you must filter the milk before chilling it. Even the cleanest milking techniques will yield an occasional hair, fleck of hay, or other contaminant. After each milking and filtering, ideally through a single-use, disposable filter, the filters should be inspected for debris and signs of

CHILLY CHANGES—HOW COLD STORAGE AFFECTS MILK QUALITY

Whether held in the refrigerator or frozen, milk—pasteurized or raw—continually changes, and not for the better. Many of the components of milk, such as the protein structures and associated minerals, are negatively affected the longer they are held at cold temperatures. In addition, cold-loving bacteria (read more about those in chapter 4) will survive and even grow over a range of cold temperatures. These bacteria are usually spoilage and pathogenic microorganisms and account for many of the flavor changes that affect raw milk during storage, thanks to enzymes they release. Goat's milk, in particular, often becomes "goaty" due in part to the damage of fat globules by bacterial enzymes. In chapter 13 I'll talk more about the best ways to freeze milk for storage.

abnormal milk, such as white flakes, crystals, or blood. If you observe debris in more than the tiniest amount, you know your udder cleaning routine is inadequate. If you see these abnormalities, you should check each animal for udder health. Depending upon the degree of material seen on the filter, you'll need to make a quality control decision about whether to keep the milk. Note that blood in the milk will pass through the filter and may be observed later as a rather heavy, brownish residue at the bottom of the milk container.

If you're hand-milking, you can pour the milk from your bucket through a stainless steel filter system outfitted with a disposable filter disc. If you use a milking machine, a small "in-line" filter can be used. These filters consist of a plastic shell containing a stainless steel coil over which a disposable filter, or sock, is placed. As the milk goes through the tubing, it passes through this filter before entering the milk can. Large dairy milking systems include a larger version of this type of filter.

After filtering, you must chill the milk rapidly. Chapter 8 covers many approaches to chilling milk, including the use of bulk tanks that have a built-in thermostat, chiller, and paddles to stir the milk while it chills. For a home-scale operation, you can set the milk pail or container in an icy water bath

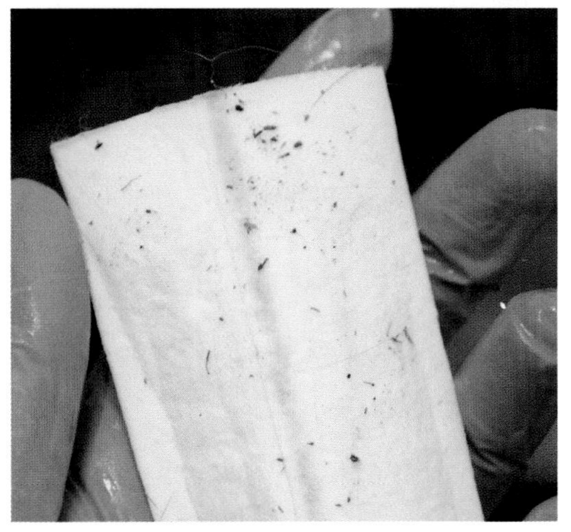

Inspecting milk filters after use is an important step in analyzing the effectiveness of the milk collection process. The filter at the top shows an acceptable level of debris. The filter at the bottom shows an unacceptable amount; better udder cleaning techniques are needed.

and stir it by hand. However you choose to chill your milk, you should monitor the temperature to ensure that your milk reaches its goal temperatures within the time limit. As I have said several times, these goals are of the utmost importance for milk quality, both for flavor and safety, and must be monitored properly.

In cases in which milk must be pumped from the parlor to a bulk tank, from the bulk tank to the pasteurizer, and so on, milk can be damaged through over agitation. In fact, more flavor flaws, especially in goat's milk, are associated with excessive agitation and the subsequent damage to the fat globules than any other factor. Fortunately, the pumping equipment of today does a much better job of gently moving milk than that available even a decade ago.

Caring for Raw Milk

Despite looking like its mass-produced commodity cousin, farm-fresh, unprocessed, unpasteurized milk has special handling requirements that reflect its living, mutable nature. A home producer can observe freshness and storage conditions easily. If you're selling or sharing your milk with people who are not a part of the daily processing, however, I recommend taking additional steps to inform them about proper handling recommendations. *Safe Handling—Consumers' Guide, Preserving the Quality of Fresh, Unprocessed Whole Milk* by Peggy Beals is a great booklet to give to herdshare customers or to have available to sell if you are retailing your milk.

I suggest pretending that raw milk is ice cream—
leave it out only as long as you would leave the frozen dessert
before it would begin to melt.

Once it's been purchased and removed from the store or farm refrigeration, raw milk should return to and remain at refrigeration temperatures until used. If you pick your milk up from the grocery store or farm, it is a fantastic idea to always have an ice chest and ice blocks along for the ride home. Clever raw-milk producers who sell milk directly from their farms might even have an "ice pack exchange" program so that you can swap your thawed ice packs for fresh cold ones. I like to suggest pretending that raw milk is ice cream. Leave it out only as long as you would leave out the frozen dessert before it would begin to melt.

All foods are prone to contamination after they reach your home. Indeed, it might be argued that pasteurized milk is even more vulnerable because of the lack of any competitive bacteria that might limit the availability of the milk to the growth of introduced bacteria, but either way, being constantly aware of what other foods you have handled is critical to food safety. In fact, the other day I was putting a pack of raw hamburger in the refrigerator and thought about its placement—where to put it? Near the raw veggies? Near the eggs? In the meat and cheese drawer? What about the pitcher of raw milk? It was so awkward, in fact, that I had an ironic, slightly sarcastic thought that if the federal agencies really cared about our safety they ought to require us to have a separate refrigerator just for raw meat and receive mandatory training before we are allowed to handle it. The point is that you always should be aware of good practices in the kitchen—hand washing, proper storage, and placement of more vulnerable foods, especially those that will be consumed raw, so that they are not contaminated by other products that are of higher risk.

**Be aware of good practices in the kitchen—
hand washing, proper storage, and placement of more vulnerable
foods, especially those that will be consumed raw.**

The cleanliness and size of the container also will influence quality. If you provide your own milk container, or transfer milk from a bulk tank or milk can to another receptacle, making sure the container is clean and sanitary is very important. If the container is glass or stainless steel, you will be able to ensure cleanliness far more easily than if it were plastic, which is easier to scratch and subsequently harder to clean thoroughly. Washing and sanitizing with boiling hot water or in a dishwasher is a great way to keep milk containers clean. Chapter 9 will cover more options for cleaning and sanitizing milk containers. The less airspace in the container, along with the fewer times it is opened and poured, will reduce the likelihood of spoilage or other microbial contamination, as well as help preserve flavor. For example, using two half-gallon containers in sequence instead of one single 1-gallon container will probably better maintain the milk's quality. A tight-fitting, rust-free, easily cleaned lid is also important.

· 6 ·

Farm Management for Superior Milk

It makes sense that the healthiest milk will come from the healthiest animals. Many factors, such as genetics, environment, and nutrition, contribute to good health. Although we often discuss these factors as they relate to human health, they're sometimes overlooked in the context of milk cows, goats, and sheep. The dairy producer, however, has significant control over all of these factors and can greatly influence not only the health of his or her animals, and therefore the animals' susceptibility to poor udder health and disease, but the flavor, nutrition, and quality of the milk as well. The healthier the animals, the better and safer their milk is likely to be. Although I won't be able to cover these topics in the depth required for mastering animal husbandry, you should come away with a greater appreciation for the lengths a top-notch farmer goes to in the care of his or her animals—and I hope you will see yourself, now or in the future, steering your farm, its animals, and its products toward optimal well-being.

Genetics

For many years, even centuries, dairy animals have been bred for making milk—as much milk as possible. Although some farmers have had the foresight to also select for other characteristics, such as sturdy, well-functioning legs and resistance to disease, they are the exception. (See the Schoch Family Farmstead Profile later in this chapter.) The modern dairy cow is, for the most part, a living machine designed to produce a maximum amount of milk for the minimum amount of feed (a concept known as feed conversion ratio). With rare exceptions she barely resembles the milk cow of our forebears, or even those from thirty-five years ago. In her book *Milk: The Surprising Story of Milk Through the Ages* (Knopf, 2008), author Anne Mendelson notes that the average yearly milk production for a dairy cow in 1975 was 10,000 pounds (about 1,176 gallons) while in 2005 it was nearly 20,000 pounds (about 2,353 gallons). Because of financial pressures the lifespan averages of cows also have decreased as farmers cull (usually meaning send to slaughter) cows whose production drops when they pass their peak milk production year, usually by age five or six.

Goats and sheep continue to predominate in parts of the world where land would be considered marginal for raising dairy cattle. These healthy goats at Black Mesa Ranch in Arizona benefit from the variety of browse available in the high desert.

As dairy cows' udders have grown, so have their bodies—with overall size increasing greatly during the last century for all of the major dairy cow breeds. Their larger bodies mean that sturdy legs and feet—which have not always kept up with body size—must bear more weight. In addition, the vast percentage of modern dairy cows spend their lives not walking to and from pastures with the gentle give of the earth and grasses beneath their hooves but in hard-floored loafing barns and paddocks. Although dairy goats and sheep have not typically met the same genetic fate, they too have suffered from the genetic "improvement" of increased milk production. Large-production dairy cows and goats are prone to hypocalcemia, also known as "milk fever," and dairy sheep are notoriously fragile at lambing time, compared to their meat-providing cousins, which have largely been bred for ease of calving, resistance to parasites, and thriftiness.

In cultures with traditions of grazing, allowing the calves to be raised—at least in the beginning—by their mothers, and without, perhaps, a ready market for animals past their peak production (usually destined to become cheap hamburger), different priorities have guided farmers. With some exceptions, where ample high-quality, affordable grazing land is available, animals that can move out across that land and convert grasses efficiently into milk tend to be more valued. Smaller-bodied cows, whose great weight does not damage the soils as much, cows with good mothering skills, and those with more resistance to internal and external

Jersey-Shorthorn cross dairy cows coming in from pasture to be milked at Stone Meadow Farm in Pennsylvania.

parasites are far more suited to this type of lifestyle than a massive, 1.25-ton cow accustomed to moving only from the milking parlor to the twenty-four-hour buffet of the total mixed ration (TMR), then back to her padded loafing stall to chew her cud. The modern dairy animal puts most of her energy into making milk, not going out to get food. In addition the contemporary dairy farmer is often forced by economics to focus on maximizing production over a shorter cow life span—rather than on mothering skills, thriftiness, and animals that are built to live decade long productive lives.

The small dairy farmer who has a market for boutique milk, whether it be raw or pasteurized, or for high-end cheese has the opportunity to choose and breed for a smaller, hardier cow. Fortunately some of the breeds less popular for massive, confined dairy settings are still quite hardy and "easy to keep" (a phrase that describes an animal that doesn't take a relatively large quantity or variety of feed to maintain a healthy body weight) compared to their extreme dairy counterparts. Some commercial dairymen with an eye to the future have even bred the ubiquitous black-and-white Holstein Friesian cow, so often the poster-cow of industrial dairying, to be smaller, hardier, and longer lived, so don't judge a cow by her color.

Environment

A good environment for a cow (or any dairy animal) is one in which physical and mental stress is low. Although it is fairly easy to appreciate what might constitute physical stress for a cow, it is not as easy to understand mental stress.

Since we humans come from a completely different perspective from our perch at the top of the food chain, we tend to attempt to respond to animal behavior as if they were also humans. One of my favorite authors and a personal hero is Dr. Temple Grandin, a brilliant and highly accomplished autistic woman with amazing perspective and candor about animal welfare. I highly recommend all of her books if you are interested in better understanding the animal psyche. Of particular pertinence is *Humane Livestock Handling.* Her approach helps people relate to how physical stress can lead to mental stress that will decrease the health and productivity of all types of livestock. In chapter 7 we'll cover some specific ideas for creating a low-stress environment for housing your small dairy herd.

Physical Well-Being

From standing on hard surfaces to drinking from unclean water troughs, there are many opportunities for physical stress in the life of a dairy animal. Let's go over a few of these and see if we can get you to start thinking like a cow!

Water Troughs

Clean water troughs have been shown to reduce the likelihood of contracting disease-causing organisms such as listeria and salmonella (who like cold, wet places) and to help prevent the spread of diseases such as Johnes (pronounced "yo-knees"), a chronic wasting disease. Research by Kim Cook, a microbiologist at the Agricultural Research Service in Bowling Green, Kentucky, showed that a good way to prevent the spread of Johnes disease on dairy farms is to use stainless steel water troughs and add an appropriate dose of chlorine to the water.

There should also be enough water troughs or stations for all animals to have access without overcrowding and bullying. In addition to the emotional stress this causes, animals are not as likely to drink enough water if they have to fight for a spot at the trough, leading to reduced body functions and lower milk production.

If the water is too cold or too warm, animals also will limit their intake, so keep their water at a comfortable temperature; this will also prevent their bodies from using extra energy to adjust to the shock of cold water (or hot water on a hot day). This can be done by locating troughs out of the direct sun in the summer, using water heaters in the winter, or using water tanks that are otherwise protected from drastic temperature changes.

Sleeping and Relaxation Areas

Just as the number and accessibility of water troughs need to reflect the number of animals in the herd, so should comfortable, well-ventilated, and safe sleeping and relaxing areas. When cows are on well-maintained pastures with shade and shelter, they can pick their own beds, but those in more confined areas will need the farmer's help in this matter. Not only will comfortable, well-bedded areas allow the animals a peaceful place to rest and ruminate (chew their cuds), it also will help them remain clean, reducing the risk of manure contamination during milking.

Footing

Any surface the animal has to walk on frequently, such as pathways to the pasture, on her way to the milking parlor, or just standing around, should be impervious both to her weight and to erosion and muckiness. It also should provide good traction so she does not slip and fall. Surfaces for cow traffic take a lot of abuse and are the most difficult to maintain, especially in wet climates and seasons. As I have mentioned, goats do not have the same issues with regard to their impact on the land. For cows with access to pasture, paved pathways that fan out to a wide gravel or other well-drained surface that encourages the gals to spread out as they make their way to pasture are a good idea. But no matter how conscientious you are about maintenance, you will encounter situations that are a challenge; for example, in most climates a certain amount of mud and muck is simply a part of winter dairy farming.

Emotional Well-Being

Sheep, goats, and cows are all animals that evolved as prey animals. They are herbivores and live in herds. They are comforted by the presence of more of their own species, usually those with which they have grown up (called a peer group). Because they evolved as prey animals, they are constantly fearful of being attacked by a predator. Humans are predators and most prey animals can sense this about us. If you can keep these two factors in mind, it will help you see our behavior through the eyes of the animal.

Herdmates and Pecking Order

Animals that live in wild herds grow up being reared by their mothers alongside others born at the same time. They form lifelong attachments to those in their peer group that help them feel comfort and safety. Although it is not always possible to keep enough animals for a natural herd to form, it is one of the best methods for decreasing stress and increasing productivity. Tactics to increase the likelihood of this dynamic include pen rearing of similar age groups and the use of so called "nanny herds" of animals that will readily nurse and raise another animal's offspring and, for various reasons, are not ideal candidates for remaining with the main herd.

Herds are hierarchies, not democracies, so you have to expect a certain amount of jockeying for position to occur periodically, even within a herd of animals that were raised together. If a new animal is introduced, tremendous upset occurs in the herd order. It all will settle out, but it can affect health and production in the meantime. As I mentioned above, groups of similar-aged animals introduced together tend to cause less disruption than a single animal.

Human Interactions

People working with milk animals often grow frustrated with what seems to be silly, irrational, and even "bad" behavior. Even if the human doesn't act on that frustration, the animals will sense it and most likely behave even worse. People

Even cows enjoy physical attention from humans. Here a young heifer gets her head scratched at Runnymeade Farm, Oregon.

working with these animals need not only to be trained to see their actions through the eyes and instincts of the animals but also need to learn to outthink the behavior and have options for accomplishing their work. For example, a cow who is being trained to milk is likely to be nervous and to want to kick. Trying to have her be first in the milking parlor is unlikely to be successful. Instead, she can be placed between two herdmates who are already comfortable with the procedure and hobbles applied before the kicking begins. Most frustrating situations can be managed if we use our bigger brains to out-think and accommodate the animal and her fear.

Overcrowding

Whether it be at water troughs, at feeders, in loafing areas (where they can all just hang out), or in open paddocks, animals need a certain amount of space to feel

HERD BIOSECURITY

Biosecurity is a fancy term for awareness and practices meant to prevent the accidental introduction of diseases, infection, and illness to your herd. In countries with histories of horrendous consequences of disease outbreaks, such as foot-and-mouth, where thousands of animals had to be destroyed, the importance of herd biosecurity is understood. The USDA provides the following components that should be a part of a good biosecurity plan:

- Establishing quarantine periods for new animals being introduced into a group or facility
- Washing and disinfecting crates or other equipment entering the facility
- Allowing only essential personnel to enter animal buildings
- Providing protective clothing and footwear for service personnel and other visitors who must enter the facility
- Securing buildings to keep out all undesirable animals, both wild and domestic
- Maintaining high standards of sanitation in animal housing areas
- Identifying and segregating sick animals, including adequate removal and disposal of dead animals.

If you notice, the first two points above are in regard to new animals entering the herd. A closed herd, in which new stock are provided by your own herd (with the occasional exception of male stock) is one of the best ways to improve herd biosecurity. If you visit a farm and are asked to sanitize your footwear, or if you return from a trip abroad and are asked at customs whether you visited a farm, don't take it personally. Animal health and lives might depend upon your honesty. By the same token, don't feel bad about enforcing your own biosecurity policies.

comfortable and unthreatened by others in the herd. Many commercial herds of cows and goats, especially, allow for only the minimum square footage per animal (in chapter 7 I'll give you some suggested minimum square-footage requirements for all dairy animals). For those farmers whose priority is milk volume, this might seem like a good idea, but the argument can be made that productivity and health then decrease by such an extent that adding more animals does not truly increase the volume of milk they produce. When using a formula for the minimum square footage for each type of animal, keep in mind that many other factors come into play that will alter the end result, such as peer groups, existing hierarchies, and the elements and weather. An observant farmer will be able to conclude over time what works for his or her herd.

Feeling Vulnerable

Cows are large enough usually to feel secure heading out to large pastures to graze without worrying about predators. But goats and sheep seem far more aware of their "snackable" size and may not venture as far from the security of their barn as cows would. In fact, I know several goat dairy farmers who don't even have their distant pasture line fenced, since the goats will never wander that far. Livestock guardian dogs, sheepdogs, and even llamas living with small stock usually provide a sense of security for the animals, however, and herds usually will follow these protectors farther afoot for grazing.

Nutrition

When I was a dairy cow 4-H leader back in the late 1970s and early '80s, I bought M. E. Ensminger's *Dairy Cattle Science.* I also talked a community college professor teaching a class on equine parasites into letting me study bovine parasites instead. While studying the book and taking the class, I realized that nutritional sciences, as well as the sciences related to parasites and how they affect the well-being of the animal, are not only vast but are constantly undergoing changes. On one hand this was unsettling, but on the other it nurtured my desire to feed and care for my cows using observation and developing intuition. Tim Wightman, the president of the Farm-to-Consumer Foundation and an educator on dairy cow management, said it best: "You have to look at the cows and they will tell you how they are doing." So just how will the cow, goat, or sheep tell you if she is thriving? Here are some things to look for:

- Cows and goats: Shiny coats (when dark colored), smooth appearance, not scruffy or patchy. Sheep: Even coat of wool for time of year.
- Clear, bright eyes; inner eyelids pink, not pale or yellow.
- Proper body weight and conditioning (learn more about body condition scoring in some of the recommended books in the appendix).
- Plenty of time spent ruminating (chewing cud).
- Normal affect—posture not hunched, ears and tail being held in normal position, alert, and behavior typical of the individual.
- Normal somatic cell counts (SCC): Monitor and determine what is normal for the herd, and respond rapidly to deviations. See chapter 10 for methods for monitoring SCC.
- Manure: Changes in manure texture are immediate indicators of problems having to do with feed changes, parasites, or other health issues.

Pasture and Forage

When you look at a green, lush pasture, it might look like optimal feed for grazing animals, but in reality it could be providing very little nutrition. For pastures to provide nutrients, they must include not only a wide variety of edible plant species but those plants must be growing in highly mineralized, microbial rich, and nutrient-available soils. In the book *Earth User's Guide to Permaculture,* author Rosemary Morrow suggests that most of us will never even see healthy soil in our lifetime. The arable (farmable) land of the earth has been worked and abused for so long that we think of it as normal.

Thankfully there are some magnificent exceptions to this. Jack Lazor of Butterworks Farm in Vermont discusses some of his approaches to land management in his new book, *The Organic Grain Grower: Small-Scale, Holistic Grain Production for the Home and Market Producer.* Not only was Butterworks Farm one of the first certified organic dairies in the United States, but Jack has spent the last thirty-five years mineralizing the soils of his 400 acres of farmland, believing

RUMINANT REASONING—UNDERSTANDING THE FOUR-PART STOMACH

Animals that chew a cud, or ruminate, are amazingly well adapted at turning fibrous plant material into usable nutrients. Their secret is a four-chambered upper digestive system, instead of the single stomach humans and most other mammals have. Although food is immediately submerged and coated with acid and enzymes in a human's stomach, in a ruminant's digestive system, the first chamber, the rumen, is basically a large fermentation chamber, with a slightly acidic pH of about 6.5 to 6.8, but is not nearly as acidic as our stomachs (which are usually less than 4.0). The animal's rumen counts on a specific profile of bacteria to ferment the food and manufacture some critical, essential vitamins.

Rumen function is so well balanced and so critical that anything that throws off its normal processes, pH, or microbial population, such as a sudden change in food, too much or not enough grain, antibiotic therapy, stress, or illness can cause a cascade of problems from acidosis and the death of essential bacteria. From there things can go downhill and even lead to death.

When baby ruminants are born, their rumens are not functioning. A special fold of skin forms a tube when the baby suckles its mom (or a bottle) that diverts the milk directly to the fourth chamber—the true stomach, or abomasum. As the baby grows and begins to nibble on fibrous material, straw, bits of hay, and grain, the rumen begins to function and develop its bacterial population. Farmers with a good hand on baby rearing will nurture rumen development; or believe that they are. Philosophies and approaches to this goal vary quite a bit.

The rumen ferments different feedstuffs in different ways. It is currently believed that grains and high-energy, high-carbohydrate feeds settle to the bottom of the rumen and reticulum (the second chamber of the cow's stomach), while long-stem fibers form a feed mat and float (see the illustration of the cross section of a ruminant's upper digestive system below). Feedstuff at the bottom of the rumen also includes the previous day's fiber that has been fermented and ruminated.

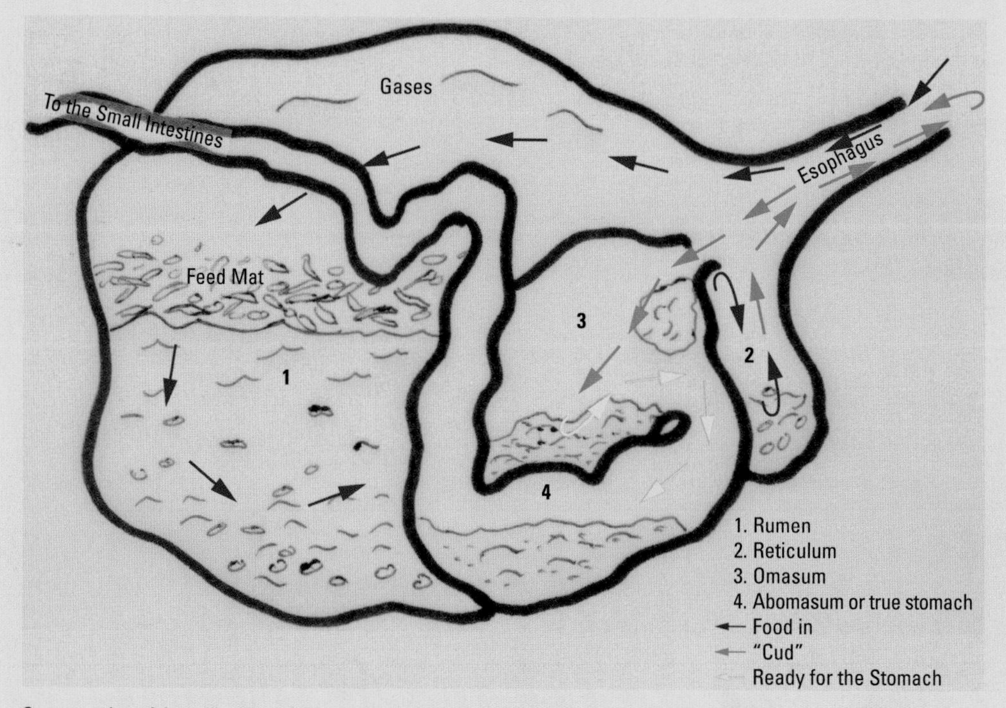

Cross section of the upper digestive system of a ruminant viewed from the animal's right side

It moves forward and up and over into the reticulum. The thinking goes that if grains are fed following more fiberous feeds that the grains can be caught in the floating, fiber feed mat and their fermentation slowed, resulting in better feed utilization, stable rumen pH, and the ideal bacterial populations. Once in the reticulum, the fermented, chewed, and broken down porridge moves to the next chamber, the omasum, and finally the well-broken-down soup—including a good portion of the microbial mass that performed its fermentation duties in the first two chambers—moves on to the last chamber, the true stomach or abomasum.

Rumen function is so critical that most veterinary medicine schools have a living cow with a permanent, well-healed opening, known as a fistula, through which can learn more about the amazing operation of the rumen.

that only through extremely nutrient-rich feeds can his cows remain healthy—and his products at their peak. For more on grazing and forage nutrition, check out Acres USA (contact info in the bibliography). Their periodical has many excellent articles, listings of workshops, and publications on this topic.

In addition to rotational grazing of paddocks, in which fields are never overtaxed and are given time to set deep roots, many farmers use multiple species in rotation to best utilize the feed height, since cows, goats, and sheep all graze the grasses and forage down to different lengths. Although goat and sheep manure—small, rounded pellets—tends to "spread itself," cow patties—big, soft plops—might need to be lightly spread. This can be accomplished in many ways, from

An overgrazed, undernourished pasture showing grass cropped extremely short and extensive growth of inedible *Anthemis cotula*, also known as "stinking chamomile."

SMALL DAIRY PROFILE: SCHOCH FAMILY FARMSTEAD, SALINAS, CALIFORNIA

The Schoch Family Farmstead's road to raw milk has been a long journey, more than seventy years to be precise. The dairy sits in one of the most naturally fertile and verdant areas of California, the Salinas Valley—also known as the "salad bowl" of the nation. Although the climate and crops are ideal for dairy cows and milk production, Schoch is one of the few remaining farms in the Monterey area—most having given way to the far more profitable enterprise of vegetable row crops and ornamental plant monocultures. But for patriarch John Schoch, son and nephew of the Swiss immigrants who founded the dairy in 1944, the love of cows runs deeper than his wallet.

When we first visited the farm in early 2012, the youngest of the Schochs' sons, Beau, took us on a tour of the farm, the dairy, and the new cheese plant that the family was just starting to build. Beau, whose "day job," as he calls it, is working for a branch of the USDA that helps farmers with developing strategies for the conservation of their land and natural resources, has a vision for the dairy's future. Along with his two brothers, Ty and Seth, and the full support of their parents, Beau has steered the farm toward a new vision of sustainability and artisanship. To close the financial gap, Beau began hauling a bit of the farm's milk, all of which is normally sold to a large dairy cooperative, to a neighboring artisan cheese facility and making delicious Schoch Family Farmstead cheese.

As we walked the farmland with Beau, he pointed out waterways he was restoring with native shrubbery along the banks and fields that had been fallow that were now lush with herbage and dotted with the farm's hundred black-and-white Holstein cows on rotational grazing, and he talked about the products they would make that potentially could save the farm. Raw milk is one of those products. The new plant, located next to the dairy, will allow the farm to expand its cheese production and utilize fluid milk and yogurt sales to create a balance between milk production, local demand, and income generation—in other words, sustainability.

Beau approaches raw-milk production with the caution and awareness that comes not only from his education and work background but also from a lifetime of dairy work. A tiny 50-gallon milk chiller (tiny when compared to the 750 gallons of milk currently harvested daily and collected in the dairy's massive stainless steel bulk tank) sits ready to collect milk from select cows as they come through the milking parlor. Beau describes this approach as "cherry picking" the best and cleanest cows. The milk will be chilled, bottled in half-gallon glass bottles, chilled again, then marketed only through farmers' markets and a few select local stores. "No middle man" is an important part of the farm's vision for quality control of this special product. In addition, the new facility will include a small lab for milk and environmental quality testing.

For this family farm raw-milk sales might be a key piece of the survival puzzle. The high demand and limited availability also mean that their community will benefit from access to their carefully and lovingly produced product. There is nothing cavalier about Beau's consideration of the risks; if the quality of Schoch Family Farmstead cheese is any indicator, local raw-milk lovers will be in for a treat. Read more at *http://schochfamilyfarm.com.*

Update: Just before I finished this book, Beau let me know that the farm's raw milk is selling well, at $7.00 a half gallon. They bottle milk twice a week and test each batch for quality. As hoped, sales are helping keep this multigenerational dairy family and farm from becoming a thing of the past.

Beau and Seth Schoch, two of the third-generation dairy farmers at Schoch Family Farmstead, California.

the use of a farming implement called a harrow to the rotation of chickens through the field to the encouragement of dung beetles. Cows don't like to graze directly around the overly green high nitrogen, and are more likely to eat parasite-larva-laden grasses that will grow around the patties; they too have their preferences.

Goats are not designed by nature to graze pastures efficiently. Their entire digestive system is designed to browse on shrubs and plants at or just above their standing eye level. In addition, goats evolved eating a wider variety of browse (it's a noun, too) than any of the other milk animals. When they have access to a wide variety of shrubs, trees, tall grasses, and what some of us call weeds, they usually can obtain almost all their nutritional needs. But this is rare, so supplementation is important. (More on that in a bit.)

Goats also are more prone to parasites if placed on pasture to graze, especially if the grasses are wet, since that is when many parasitic larva climb up the plant in the hope of being given a ride to a new home thanks to a passing herbivore. Fields can be planted with taller, more shrublike forages such as chicory, sunflowers, and legumes. (Again, this section is not an extensive or complete guide to growing crops and browsing goats, so please refer to the appendix A and the bibliography for more titles and websites that will be of use.)

Although goats have a reputation for being able to eat "anything" (even the proverbial tin can, which of course they can't and don't eat), they prefer top-quality, clean feeds. Feed that has fallen on the ground, is too stemmy, is moldy, or is not high quality is always passed over for something that they deem better—when they are given the option. This is one reason they also have a reputation for being easy to care for, since they avoid dirty feed that might contain parasites, usually know when plants might be toxic, and do what they need to do to survive (such as utilize feeds that other livestock cannot).

ARE GRASSFED COWS IMMUNE FROM *E. COLI* 0157:H7?

For quite some time there has been a rural legend that cows fed no grain and only pasture cannot harbor the pathogenic type of *E. coli* and so cannot be a source for contaminating milk with this frightening, often deadly microbe. I recently have heard some serious, salt-of-the-earth farmers repeat this myth. There are conflicting studies surrounding the connection between *E. coli* 0157:H7 and a grain-heavy diet. Suffice it to say, none is conclusive. Some simple facts, however, can be applied to the topic. First, any cow, or animal, including us, can host this variant of *E. coli* in our systems. Even if we aren't grainfed. But the farther any animal's diet is from natural—and for cows a heavy grain diet is quite unnatural—the more likely it is that their digestive and metabolic systems will not be operating optimally. When health is not optimal, the immune system is not optimal, and everyone is more vulnerable to the colonization of undesirable species of microbes. Cows that are healthy can serve as reservoirs of pathogens because of previous exposure. In chapter 2 I talked about the importance of buying stock from known sources, testing new animals, and the quarantine of new animals. Over the long term, a closed herd, in which no new females enter the ranks, is one of the best ways to prevent the introduction of animals carrying disease or unwanted microbes.

Sheep are efficient grazers, with lips that are able to crop grasses selectively and closely, so it is easy for them to overgraze pastures. In general, sheep do well on the same pasture types as cows and goats do, including some browse (small brush or tall forbs) options. They are versatile and efficient. Some data, however, link decreased fertility in sheep, and other livestock as well, to the phytoestrogens (plant origin hormones) in grazed, fresh legumes such as alfalfa and clover. In addition, fresh perennial legumes (such as alfalfa and clover) carry a risk of bloat if not introduced carefully. After the first frost, this risk goes away.

Fresh Fodder

The term "fodder" typically refers to any feed for animals that is harvested and brought to them. A relatively new approach to fresh fodder involves the sprouting and growing of grass seed (grains) to only a short height, then feeding it to livestock. The seeds are sprouted and grown in trays fed by an enriched watering system. The grasses form a mat of roots and green fodder that is fed in its entirety to the animals. The fresh fodder system allows for the intensive growing of a large amount of what most believe to be a high-quality feed in a relatively small space, using racking systems, climate control, and hydroponics to grow the crops. Practitioners of this system have sprouted such seeds as alfalfa, clover, rye, barley, and wheat and sometimes refer to them as "superfoods" thanks to their extremely high nutritional value and digestibility. Although commercial systems are expensive, several creative farmers have devised their own homemade versions. Whether these systems and approaches stand the test of time is yet to be seen.

Dry Fodder

Grass, legume, and grain hays are the typical dry fodders fed to dairy animals, with legume hay such as alfalfa being at the top of the list for providing protein and minerals for increased milk production. Hay quality is a direct reflection of

IS THERE SUCH A THING AS TOO MUCH PROTEIN?

Many farmers select dairy feeds, including dry fodder and concentrates, for their high protein content, believing that this will increase production and milk protein content. However, when an animal's diet is too high in protein, its body converts it to urea, a nonprotein nitrogen molecule. Urea levels can be measured in the blood or milk as BUN and MUN. These levels fluctuate greatly based on not only feed consumption but also on water intake (which dilutes the levels) and milk production and milking intervals. When an animal produces too much urea, it affects milk quality, and the nitrogen output in solid and liquid waste products increases. The animal's feed processing and kidney function also are stressed while dealing with the excess protein.

Optimal MUN levels for dairy cows are well researched and documented at 10 to 14 milligrams per deciliter of milk. Dairy goat levels, however, are not well researched, but small samplings of several goat dairy herds indicate that levels of 20 to 30 milligrams per deciliter of milk may be normal. Increased nitrogen in waste material also can challenge waste- and land-management systems. Read more about MUN at *http://puyallup.wsu.edu/dairy/nutrient-management/data/publications/munfinal.pdf.*

the soil quality, harvest methods, and storage conditions. Hay must be grown in nutrient-dense soils and harvested at just the right time to capture as much of those nutrients as possible in the leaves, stems, and seeds (for grain hays) of the hay. Farmers who grow hay are often at the whim of the weather, however, for harvesting. This is one reason that some farmers choose to cut the fields for haylage instead (more on that next). Once harvested, hay must be stored in a way to limit deterioration as much as possible; many vitamins are unstable and will decrease as the hay ages. Protecting hay from sunlight, moisture, and dust is also imperative.

Cows eating fresh salad greens as fodder at Schoch Dairy in Salinas, California.

Silage and Haylage

To some, silage has a negative ring to it. We small farmers might think of it as fed only to cows in large, confined industrial dairies. But that doesn't mean it cannot be part of an animal's diet when it's properly managed and monitored. So what's the difference between silage and haylage? First, we all know what silos are: those stately, tall structures on old farms—now usually empty. They were built to store grains and crops such as corn, that were harvested at just the right time (often including the whole plant: corn cobs and stalks), chopped, and piled inside the silo (called "being ensiled"), then allowed to ferment. Think of it as kraut for cows. Nutritious, preserved, and easier to digest. Silage was a way for the farmer to preserve an otherwise difficult-to-store crop and provide nutrition throughout the winter. Haylage is a relatively new term for forage crops (such as grass and alfalfa hay) preserved in a similar fashion, often in bales or bags shrink-wrapped in plastic (remember that fermentation can occur without the presence of oxygen).

A lot of silage now is used on confinement dairies as a part of a total mixed ration or TMR. Corn (the entire plant), soybeans, and alfalfa are the typical blend. To this, a farmer—or, more likely, a dairy nutrition specialist who has done the research and believes he knows the exact nutrient requirements for the herd—might add other supplements. This is the section in Ensminger's book, *Dairy Cattle Science*, which I mentioned earlier, that was, for me, simply incomprehensible. Formulas, feed conversion rates, yield, and more. I now believe, as I do for human nutrition, that the nutritional science is far from complete. I approach herd management by attempting to come close to a life that nature would provide. But you must also remember, that many of these animals, changed as they are from their ancestors, would not survive well in nature, at least in the beginning.

Silage's bad reputation comes from two main sources: first, because it is easy for farmers to incorporate lower-quality feeds and questionable ingredients such as chicken manure (though this is rarely used today) and blood meal from other cows (this is no longer used because of the scare over bovine spongiform encephalitis, BSE, also known as mad cow disease). The other primary black spot on silage's reputation has to do with quality-control issues during storage—molds; the wrong type of bacteria, some of which transmit to the milk and cause problems for cheesemaking; and bacteria such as clostridium are amongst the challenges facing farmers who use silage. As with any feedstuff, however, good management makes the difference.

Grains and Energy Feeds

Grains are basically the seeds from grasses. Grains are typically higher in sugars (carbohydrates) and are a source of energy (heat) for the animal. They also provide a different balance of minerals and vitamins than do dry fodder and forage. Although feeding grain has been blamed for many of the cattle industry's problems, it is really due to extremes, not because grain itself is inherently bad for cows, goats, and sheep. Today's dairy cows and goats have been bred to put so much energy into producing milk that many of them would grow too thin without the addition of some carbohydrates to their diet. As pasture, forage, and fodder quality increases, however, many farmers have been able to reduce the intake of grains to a minimal, as-needed level. Here I encourage you, once again, to develop your skills of observation regarding body conditioning and general health. Learn more about rumen function (there is a bit about this in the sidebar **"Ruminant Reasoning,"** earlier this chapter), and see how your animals can adapt to less grain, should that be your goal.

Minerals and Supplements

I think it is fair to say that almost all dairy animals will need supplementation of minerals, vitamins, and micronutrients. Not only do animals rarely have year-round access to food that meets all their nutritional requirements, but many feeds are grown on lands bereft of minerals essential for optimal health. Different species also have different needs, so much so that feed that is perfect for one species might be deficient for another. Water too might contain minerals that help or hinder the absorption of other nutrients. It all gets pretty complicated!

There are two approaches to supplementation: The first is what I call the "anticipate and integrate" approach, in which the nutrients are analyzed, then included in the main feed so the animals ingest them along with their regular diet. The second is the free-choice approach, in which high-quality minerals and other nutrients are made available for the animal to eat as needed; you can probably guess which method I am a fan of, right? Mineral blocks, or licks; tubs of loose, granulated minerals; buffers, such as baking soda; and salts are just some of the many options for the dairy farmer and their animals.

Minerals for goats often are provided in loose, granular form, since goats prefer to not lick a block after another goat—they are surprisingly picky about the saliva

WHAT ABOUT BREWER'S AND DISTILLERY GRAINS?

In chapter 1 you read about the horrendous conditions endured by dairy cows living in the late nineteenth and early twentieth centuries in cities where dairies partnered with distilleries to produce cheap milk, no matter the cost to the cows or the milk quality. Often called "swill dairies" for the sloppy, wet mash of grains and liquid that the poor beasts were fed, there was nothing positive about anything these animals had to endure or the low-quality, contaminated milk they produced. The ultimate price was paid by the innocent infants and children who died or were sickened by this milk. In many ways raw milk's reputation is still tarnished by this approach to dairying. But it was not spent grains alone that caused the health problems for the cows; it was the lack of other much-needed feeds, poor housing conditions, and the dairies' determination to produce as much milk as possible for the least amount of money.

Here at our farm we regularly feed a small amount of wet brewer's grain (also known as WBG and spent brewer's grain) to our goats. This is a different type of grain, produced during the making of beer, from the grain left over from making something such as whiskey. I do know of at least one small dairy farmer who has integrated a certain amount of distillery grain successfully into his feeding program. The key for success in using WBG is observation and management. WBGs are high in protein (with some data saying they are over 23 percent protein), low in carbohydrates (the opposite of regular grain) because the sugars have been fermented during the brewing process, and high in fiber. They also contain brewer's yeasts that are helpful microbes for rumen function. Because they are wet, however, they are prone to molds and spoilage and must be managed properly and observed before feeding. If they show signs of mold or discoloration or stop smelling like cooked cereal, we send them to the compost pile. The key to feeding any of these odd, less-than-traditional feeds is to use them as an augmentation, not a replacement, and to monitor their quality strictly.

Fresh brewer's grain, also called spent brewer's grain, from a small artisan beer company provides an economical, flavorful, and healthy component to our goats' diet.

of others! Cows and sheep are usually a little less finicky and do well with large, solid mineral blocks. Often these blocks contain other supplements or are labeled for multiple species. Be sure to spend time looking at the ingredients and recommendations from sources other than the company marketing a particular product. Most companies make their product labels, along with nutritional information, available online. You can compare them and also consult with a local forage specialist (often available through university Extension programs) as to likely feed deficiencies and concerns in your region.

Probiotic granules, powder, or paste are used by some farmers to help with rumen function when an animal is sick or being treated with antibiotics. They also can be used as a regular part of the ration, either sprinkled on feed or mixed in with loose minerals. Studies have shown that regular use of these beneficial bacteria in livestock is showing some amazing results in preventing infection and illness from pathogenic bacteria. There is speculation that they could help reduce the shedding of bad bacteria such as *E. coli* 0157:H7 in dairy cows, which could help reduce the likelihood of contamination in milk. If you decide to add probiotics to your animal's diet, follow storage instructions and shelf life recommendations for the product, since bacteria provided in these products can easily die and become useless unless properly cared for.

Supplements that assist in maintaining the ideal rumen pH (we talked about this earlier in the chapter) are an important part of dairy animal management. Both yeast supplementation and buffering agents (which help animals maintain the right acid level in their rumens), such as calcium carbonate and sodium bicarbonate (commonly known as baking soda), are extremely important to have available at all times for ruminants. Cows often have buffering agents added directly to their ration; goats and sheep more often are offered these supplements freely, eating them as needed. Yeast, usually a concentrated brewer's yeast packaged and sold for livestock (I like Diamond XPC), does not directly buffer the rumen pH but instead promotes fermentation and microbe activity, decreasing the need for buffers. Direct buffers, such as baking soda, immediately raise the pH level of the rumen—even if it is already at optimal level—but if offered free choice, the animal can utilize it as needed.

Most animals, however, know when they need to have a dose of buffers and will eat them as needed if they are available. Fibrous feeds also encourage additional rumination time, which produces more saliva and consequently more natural buffers. It is important to remember that no amount of added buffers and supplements will fully compensate for a poorly designed or poorly managed diet.

PART III

DESIGNING AND FURBISHING THE MODEL DAIRY

· 7 ·

Housing and Habitat

Let's get down to the specifics of providing shelter, habitat, and facilities that will support the goals of healthy, content animals and efficient, effective milk collection. Remember that there is no single perfect dairy barn, either on paper or in reality. Most small farmers work with limitations such as geography, topography, regulations, existing infrastructure, and, of course, budget. But it's quite possible to adapt many situations into successful small-scale dairies.

Barns and Shelters

A good barn will provide ample room for all of its inhabitants to dine, drink, ruminate, and sleep; be well protected from weather extremes; have comfortable resting areas; and be easy to clean. A dairy barn also should be designed in a way that will make moving milking animals to and from the milking parlor as easy as possible.

LIVESTOCK COHABITATION

One important consideration that people often overlook on a small, sustainable farm is cohabitation of dairy animals with other livestock. Although a field dotted with various farm animals brings to mind a wholesome, back-to-the-land sensibility, there are many farm animals that can increase the risk of pathogen contamination in your milk. It is important to segregate work that involves other animals on the farm, such as chickens and pigs, until after dairy work is done to reduce the chances of introducing unwanted contaminants into the milk. But you should also think about keeping these same animals completely separate from the dairy portion of your farm. If your dairy is inspected, the presence of such livestock in the dairy barn likely will be a violation.

Many farmers keep hogs to consume extra vegetables, milk, and whey, but they absolutely should not share a pen with milking cows, goats, or sheep. Poultry, too, are a prime source of potentially disease-causing microbes such as campylobacter and salmonella. This doesn't mean they can't play a role in insect management that will benefit a dairy farm, as well as the many other things they add to any farm, but when milk is involved, extra considerations must be made.

Barns and shelters need to provide adequate square footage for all the animals being housed without overcrowding. Opinions vary on how many minimum square feet per animal should be provided. Several organizations provide humane certification to farms that meet their standards. Typically these parameters will give space requirements by species and age. Tim Wightman, president of the Farm-to-Consumer Foundation and animal-management specialist, offers the formula of 60 square feet for every 100 pounds of livestock. As a frame of reference, that would be about a 30′ × 20′ space for a large cow or seven or so goats. As I mentioned in the last chapter, other factors will make these formulas more or less pertinent. For example, a herd of animals that were brought together as adults—not raised as peers—will need more room until a stable pecking order is established.

These spaces need to be protected from all types of weather so that no matter the conditions, the animals can be comfortable and content. Most livestock handle cold extremes better than they do heat—as long as they can be dry and out of the wind. Owners, however, may appreciate some heat in the barn, but it is not necessary for most animals. In very cold temperatures, the deep-litter bedding method (where manure and bedding are allowed to build up under a clean layer of straw, creating heat) or thick bedding material can help animals conserve body heat and remain comfortable and productive. High humidity combined with heat typically will reduce productivity and animal comfort. Large, industrial barn fans can be helpful if such conditions will be a regular occurrence. In regions with high heat and low humidity, mister systems can help cool animals. If you live in an area with deep snows, plan on animals staying indoors most of the winter season. In some parts of this country and in Europe cows are still wintered in tie-stall barns where they are milked, fed, and sleep in place.

Small cow barns might include free-access loafing stalls that are bedded with organic material such as wood shavings, a rubber stall mat, or even a "cow mattress." These stalls are wide enough to allow the cow to lie down, but not turn around. The length is calculated so that when the cow stands and poops or pees the waste will be collected in a drainage trough just outside of the stall. Even a small barn can include such spaces as a way to provide cow comfort and cleanliness. Goats and sheep will turn around in the smallest of places, so this sort of stall system will not work for small ruminants. Goats do prefer slightly sheltered and elevated areas for resting, so spaces such as platforms and shelves can be built that increase lounging square footage. These shelves can be covered with rubber mats for increased comfort and reduced bedding needs.

For the farmer's sake these areas should be easy to access for cleaning and maintaining. If pens are difficult and awkward to clean, these tasks will become easy to overlook and put off until later, leading to dirty bedding and dirty animals. Large facilities often use a great deal of water to purge concrete pens of waste material or a gutter system and automated manure scrapers, but these are unlikely to be affordable for the small farmer. Instead, make large spaces accessible to a small tractor and smaller spaces to a large cart. Try to plan so that compost or manure collection piles are downhill from the barn so that less effort is required to transport heavy loads. An elevated slatted wood floor with a collection system beneath can be used for goats and sheep, greatly reducing bedding costs. No matter how well you design a barn, though, plan on a lot of cleaning. A tidy dairy requires a lot of dedication to the art of mucking!

Animals should be able to move between feeding, resting, and holding areas without the stress of things such as bottlenecks or blind corners where a more dominant animal in the herd might take the opportunity to bully. You also should be able to move animals through these traffic areas without too much difficulty to yourself—such as having awkward gates to latch, no way in and out that is designed for a human, or ways for reluctant animals to avoid being moved along with the herd. As with the resting areas, traffic areas should be easy to clean and maintain. Lameness in cows often can be attributed to poor-quality traffic surfaces where slipping and falling is likely.

To move animals efficiently in and out of the milking parlor, you might need holding pens. Once milked, animals should be able to return to the main housing or loafing area. The holding pen should be large enough to hold the maximum number of animals that you will be milking without their being overcrowded at feeders and while they wait. If there isn't feed to distract the animals, bullying is more likely to occur. Animals coming into the parlor stressed will not milk as well and are more likely to become injured while trying to avoid another higher up in herd order. Holding pens should, of course, be easy to clean—and be cleaned frequently so the animals will be as clean as possible when they enter the parlor.

AIR QUALITY

The air in both the animal areas and the people areas of the barn should be fresh and well ventilated. The buildup of odors from ammonia and manure is unhealthy for the animals (leading to respiratory issues and health stress, in general) but also can affect milk quality, especially when open pails are being used. Keep in mind that most healthy dairy animals tolerate cold better than poor ventilation. When we first built our barn, it did not have adequate ventilation. The removal of the upper portion of the walls helped, but overhead ventilators would have been the best choice. Well-ventilated pens also will reduce bedding needs, helping the farmer save both money and labor. The Dairy Practices Council, www.dairypc.org, offers many booklets on ventilation and dairy barn design.

Paddocks and Pasture

Animals need exercise and foraging areas to maintain optimal health. Dry paddocks—where no food is available—can provide some activity, but ideally there will be larger pasture and forage areas for your herd. Often as not, the design of pastures and paddocks will be constrained by the layout of the land, zoning laws, and other land-use restrictions. Properly managing both large and small acreage is an extensive challenge, one that I can only touch upon in this book. Before you sink a single fence post in the ground, I highly recommend consulting with experts, including the USDA's Natural Resources Conservation Service. This department works independently of USDA regulatory offices and provides incentive and resources to those seeking to optimize their land use and properly manage livestock waste (see appendix A for contact information). Let's go over some of the major considerations regarding pastures and paddocks.

Stocking Rates

This term refers to the number of animals that a section of land can support without damage to forage, soil, and features such as waterways and hillsides. As you might imagine, the stocking rate will change seasonally—as forages become more or less available and the land more or less vulnerable to hoof damage. Stocking rates also vary by species and breed as well as the management goals of the producer. Again, this is a topic that deserves input from other experts. If your state has an active land grant university Extension service, they might offer access to a forage specialist. Stocking rates can be adjusted as time goes by; be sure to pay attention to how the land responds to given management decisions.

Fencing

If I could give one thing to every farmer, it would be a magic wand that would build and maintain fences. Fencing is expensive and rarely finished. Both goats and cows can be quite challenging to fences while sheep are relatively gentle on fences by comparison. When I was growing up, our neighbors' cows, whose fencing was not quite as sturdy as our own, were frequently escaping and turning up near our pasture. The old phrase, "The grass is always greener on the other side of the fence" was probably written by a farmer or rancher watching Bessie lower her heavy neck over a section of sagging wire, long, rough tongue stretching and reaching for the perfect blade of forbidden grass. Trust me, it doesn't take much

At Schoch Family Farm several types of fencing assist with rotational grazing—the farm's original wooden fencing, metal "T" post and wire fencing, and electric net fencing.

for a half-ton heifer to pop staples out of fence posts when she wants what is on the other side. Goats, on the other hand, challenge fences by placing their hooves on the wires, poking their heads through openings, and dragging their flanks along long lengths of wire for a good scratch. They are also good at wriggling through gaps and getting stuck in sections that would keep most livestock contained.

These dairy sheep enjoy both shade and water thanks to a large live oak tree and a water trough positioned underneath.

If you will be doing rotational grazing, then temporary, movable fencing is ideal. Two or more strands of well-charged electric fencing is usually all that is needed to contain cows (as long as they are not seeking the affection of a neighboring bull—or vice versa). Goats and sheep will need electric netting (see resources in appendix A). Depending upon how wet your soil is and how motivated your goats are, however, even electric netting may not work well for these determined and wily animals. Remember that electric fencing will deliver a proper shock only if the subject is touching the fence and also has good contact with the earth (the fencing system includes a conducting rod that is driven deep into the soil).

Goats and sheep are much lighter bodied, and their hooves do not cover as much surface soil as do those of a cow. These two factors make it less likely that even the best-charged fence will deliver a properly motivating zap to the potential escapee. You can train them, though, to respect the fencing by introducing them to it while the ground is wet and the fence is well charged. Make sure that the netting is sized so that animals do not stick their heads through a hole, then get zapped, as they will likely panic and run forward, becoming more entangled. Be on hand during the first introduction to this type of fencing just in case an animal becomes snarled or decides to push the entire thing over by placing its hooves on a vertical, noncharged fence stake. I find the best way to ensure that they don't push it over or become caught up is simply by making sure that all of the good forage is inside the fence. Simple goat psychology.

For smaller-area rotational fencing, many farmers like to use hog or cattle panels. This type of fencing comes in 16-foot sections and heights from 32 inches up. The panels are made of sturdy, steel rods spaced to discourage animals from pushing through at the bottom. The panels can be attached to vertical posts such as steel rebar or other material driven into the ground. Although costly, the panels are easy to move, are virtually indestructible, last indefinitely, and can be recycled down the road—unlike most modern electric fencing. They also have the advantage in heavy brush or plant growth, where contact with the plant matter and an electric fence would cause the fence to lose its shock capability.

Fences that need to contain more aggressive animals, such as bucks, rams, and bulls—in other words, the males—need to be designed to take more abuse than those that will contain the female population. Not only are these important members of the herd larger and rougher, but they usually are motivated and determined to provide their services to a female in estrus. Sturdy, long-lasting fencing will go a long way toward keeping the breeding program under the farmer's control.

When building and designing gates, think of the convenience of the latch for you—can it be easily unfastened, opened, and closed if you are in a hurry, carrying a load, or trying to wrangle an animal in or out of the pen? I am a big fan of antechambers—a small pen just inside the main gate—that allows you to enter the gate, then enter the pen. The chamber is basically a catch pen that prevents animals from completely escaping should they make it through one gate. Ideally, gates should swing into the pen, giving you a way to push animals out of your way, rather than their pushing the gate open onto you. (I'll bet you can tell I work with wily goats.)

Climate Protection and Water

Animals are unlikely to range far, or linger long, if their trek to the back forty does not include a watering hole (most likely in the form of a galvanized or rubber trough) and a comfortable place to ruminate in the shade. If your goal is to get them out there to eat, and not have them burning all their calories ambling back and forth for water and rest, providing these comfort items is important. Planting shade trees, protecting existing trees, or building simple field shelters will supply the needed siesta station. Drinking water might need to be piped, hauled, or pumped. (The iconic western water trough with windmill-supplied water to cattle far from any electricity? That was renewable energy before the term was even coined.) Some farmers also supply a salt and mineral block or feeder near the water, since the animals appreciate, and need, a little seasoning with their meals.

Egress

As animals come and go from the pastures, they will make a big impact on the land. The heavier the animal and the more often they traverse the track, the more the damage. When moisture is present and the earth is wet, this can create quite a muddy mess. Not only is it hard on the land but animals will be dirtier coming into the milking parlor. Although goats avoid rain, mud puddles, and water and usually will not range when the weather is inclement and the soil soft, their bovine sisters are impervious to all but the worst that Mother Nature has to offer. Protecting pathways from the damage of heavy hooves is a challenge—but one that you can anticipate. Though expensive, paving grids can be especially useful; these are heavy plastic grids that are filled with gravel or other drainage material and that allow for drainage while holding the soil beneath in place.

When I was a kid, our barn had one route for the cows to come into the barn from the pasture. During the winter this pathway was what we call a "boot sucker," literally pulling your rubber boots off as you tried to walk through the deep muck. There was a lot of one-foot balancing as we tried to retrieve the trapped footwear. It was one of the memories that led me to goats.

Feed, Water, and Supplement Stations

Although building a proper barn and dairy is usually a one-time expense, plus some maintenance, feed is an ongoing, often escalating, cost for the farmer. Protecting feed during storage and from waste while feeding will help both animal welfare and the health of your pocketbook. Let's go over feed storage and transport; feeders; and water and supplement stations.

Whether it is hay, haylage, grain, pellets, or other feed, all foods are subject to deterioration, even when they're properly stored. Light, moisture, heat, and time are some of the factors that will make the ton of hay you purchase in August very different from the same hay at the end of winter, even if you store it under optimal conditions. Although I cannot address every condition that your feed may face during storage, let's take a look at where otherwise unprotected feed (that is, not in waterproof bags or wraps), such as baled hay, will be stored and then assess the amount of sun, wind, and rain exposure that the storage location is likely to experience. The opening to our hay barn, which can hold at least 50 tons of hay and straw, faces away from the usual direction of inclement weather and is shaded during the brightest part of the day. Even so, we hang heavy tarps across the front and still experience some loss from ceiling condensation during the winter.

Stored fodder should be assessed at each feeding for mildew, mold, dust, and debris that might have contaminated the bale during harvest. You also should anticipate vitamin loss and provide other ways for the animals to get these vitamins. I supplement our stock with a source of vitamin A, usually pumpkins we grow, during January to make up for the natural loss of this vitamin in the dry fodder, since vitamin A deficiency is associated with chronic conjunctivitis (also known as pinkeye).

You'll also want to move feed to be closer to animals during poor-weather seasons. Although it won't hurt for dry fodder to get a little wet when animals are about to consume it, it is frustrating to have to push a cart through mud and muck or heavy snow. Although you can't foresee every situation—and even if you could, it is unlikely that you can be prepared for all the possibilities—if there are opportunities to make feeding time less burdensome and more efficient, you will never regret addressing them early.

Feeders should be easy to clean and built to discourage animals from placing their hooves inside. And they should be situated so that manure cannot be easily deposited where hay should be. It is fairly easy to keep cows' feet out of feeders, but not goats'. On the other hand, goats don't have the ability to project their

These mineral and buffer tubs and water trough at our farm are easy to inspect and refill, and they stay cleaner than if they were located within the enclosure.

manure quite as far as a cow does. If the young are in with the adults, it is virtually impossible to keep lambs and kids out of the feeders, so making sure they cannot be harmed if they use Mom's dinner plate as a jungle gym—as well as more frequent inspections and cleaning—may be your best option.

In chapter 6 I talked about the importance of adequate numbers of water troughs for the sake of reducing physical stress. In addition to number and easy access, troughs should be easy to clean. Not only will animals limit their intake if their water become dirty, but clean water is an important way to limit the animals' intake of unwanted bacteria and other pathogens that in the best case increase their stress and in the worst case cause illness and possible transference to milk. By the same token, supplement stations should be easy to check, refill, and keep clean. Why pay for and provide valuable nutritional opportunities only to limit intake and throw out a lot of product?

Management during Procedures

As you design or plan modifications to your facility, remember that animals will get sick and need other routine procedures that might require restraint. A small, well-handled herd might be tame enough that a simple halter, collar, or head stanchion will do, but in a herd in which most of the stock have been raised by their mothers or otherwise not handled much by humans, other equipment and spaces will be useful. A squeeze chute for cows and a catch pen for any reluctant animals will help move the procedures along with less stress for all.

I also recommend a quarantine or sick pen located close enough to the rest of the animals that the segregated animal is not stressed by isolation, but farther than "sneezing" distance to prevent the spread of airborne transmitted illnesses, as well as any direct contact. If it is not possible to locate the pen close enough, you can consider placing another "sacrificial" animal in with the new herd member—one whose possible exposure to illness, for the sake of the entire herd, is an acceptable risk. This pen will be useful for separating new animals during their quarantine period and an ill animal from the herd. Often they will benefit from separation, not necessarily for the prevention of the spread of an illness but for their own tranquility while they recover.

Offspring Management

One of the most labor-intensive times of the year is birthing season. On some farms the arrival of new babies is a multiseasonal event. Having facilities for proper care of expectant mothers as well as for rearing young stock will help make an otherwise exhausting and difficult time far more manageable. There are different ways to approach maternity and livestock rearing, from a hands-off, raised-by-Mom and with the herd approach, called dam raising, to removing the babies immediately after birth to be raised by the farmer. Over the years I have come to feel that no one approach is right for everyone. In the bibliography I recommend specific animal care books for goats, sheep, and cows. As with all management practices, you may need to try several different approaches until you find the one right for you, your herd, and your lifestyle.

If you will be dam-raising babies, the most important concern is ensuring that your paddocks and fields are "childproof." Areas should be assessed for hazards such as low-hanging water buckets and troughs—a drowning waiting to happen; fences in need of repair where offspring might squeeze through—or small predators squeeze in; feeders that young can climb in and dirty with hooves and excrement; and filthy conditions leading to illnesses in the young such as navel ill or bacterial intestinal infections.

If you plan on trying to observe each delivery, maternity pens are helpful features. Some expectant mothers will experience behavioral changes as delivery nears. Maternity pens shared by several mothers, even those who are normally herdmates, might be a setting for unnecessary stress from jousting and posturing. If this is likely, having solitary birthing pens can be helpful. Even when isolated, however, herd animals such as sheep, cows, and goats do prefer to be able to see the rest of the herd and might become agitated and restless if they cannot.

Having clean, sheltered maternity areas helps with reducing stress on the mothers, ensuring a healthy start for the young, and allows the farmer to quickly assess the progress of labor.

Individual hutch rearing of young calves, although not considered humane by some, often is practiced to help the farmer manage health and feeding. Here a young Jersey calf is being raised individually in a typical small wooden stall inside an old barn.

Communal rearing of small groups of calves can provide social stimulation and adequate room for exercise. If they aren't being raised in groups or by their mothers, though, young calves are often still raised in small, individual pens sheltered by a structure called a calf hutch. Hutch or stall rearing came about as herd size increased and when many dairy calves were raised for veal—a practice that encouraged immobility to keep the meat of the milk-fed calf at its most tender.

Now, although plenty of farmers still direct the lives of many young animals toward our dinner plates, it usually is done with the basic needs of the animal in mind. Hutch or stall raising still is promoted, often by those marketing hutches, as the best way to limit the spread of typical youngster health problems such as pneumonia—and they are right. However, this concern can be offset through other interventions that boost health and resistance to illness—but not without added labor, observation, and cost—something the larger farmer may not be able to afford. Goat kids and sheep kids are more likely to be raised with their moms or in groups of similar ages.

If you are raising young apart from their mothers, you can feed them milk or milk replacer by individual bottles, pails, or bucket feeders (also known as lamb bars). No matter what you decide to feed the calves, lambs, and kids, having a dedicated space for mixing their feed, storing milk that is not intended to be used for the humans, cleaning baby bottles, and storing other baby supplies is one of the most invaluable spaces on a dairy. Unless you are producing milk only for your own use and in a home kitchen, having a space that serves this purpose helps the work flow and reduces the possibility of cross contamination. Activities such as cleaning bucket feeders, dosing medications, and mixing milk for young ideally should not occur in the same space that milk for humans is processed.

When this priority finally dawned on me, I began advising such a room on dairy projects on which I was consulting. After visiting one of these farms and seeing their baby milk room, we stole some space at our own farm and built what we dubbed the Kid Milk Kitchen, or KMK for short. Not only did this make our spring kidding season function more smoothly, but it also freed up space in our milk house (more about the milk house in chapter 9), making that space function more efficiently.

· 8 ·

The Milking Parlor

Whether you are milking one goat or fifteen cows, the milking parlor is the keystone of the dairy. In the last chapter you learned how to ensure that your milking animals have the proper facility to be as clean and healthy as possible. In this chapter I'll walk you through all of the things you should consider before designing and outfitting a milking parlor, even if you are a single-cow dairy. Over the years our milking area has gone from the porch of a shed to inside a three-sided shelter to inside a horse trailer while we were between residences to inside a converted garage with a wooden, four-goat milk stand that I built and finally to our current Oregon Grade A parlor. I know the importance of flexibility and the need to adapt to situations, both financial and logistical. That being said, for this book let's focus on the ideal milking parlor.

We'll start with tips for making the best decisions about construction materials, then cover milking equipment choices and milk-production monitoring options. If you are considering building a licensed facility, many rules and regulations will need to be addressed. In chapter 2 I covered many of the steps that you may encounter during the process of building a dairy that will meet regulatory requirements. In this chapter we'll cover the specific standards as defined by the Food and Drug Administration's Pasteurized Milk Ordinance (PMO), the federal standards regarding Grade A milk production. Keep in mind that each state may have its own set of standards that deviate from the federal guidelines. Don't be surprised if a model facility in one state looks very different from its counterpart in another state! If a product is produced using only ingredients from within the state and also is sold only within the same state, federal regulations do not apply.

Construction Tips and Considerations

The milking parlor is the first time that the otherwise sterile milk in the udder will be exposed to the environment—through open-air milking or through air that enters the milking equipment during machine milking. It is your first opportunity either to keep the milk as the animal created it or allow airborne bacteria,

This tidy, small-scale goat parlor at Black Mesa Ranch in Arizona is set up to milk two goats at a time, while two others await their turn.

yeasts, molds, and microscopic debris to contaminate the pristine liquid. Even if you follow all the procedures for clean udders and teats, the air in the parlor must be clean as well.

To accomplish this you must be able to thoroughly clean any surface in the milking parlor that contributes to the clean collection of milk. Ideally, ceilings and walls will be washable, since they collect dust and debris that can decrease air quality and contaminate the milk. By the same token, the milk stand or platform should be made of material that is easy to scrub and keep clean, and the floors should be made of concrete or other equally durable and cleanable material that can be washed down routinely. Don't forget to provide good lighting so you can see if both the animals and their environment are clean. And finally, you need to develop a cleaning and maintenance routine, then follow it.

A sink that is dedicated to handwashing is a super idea; this way, when you come in from other chores, you can wash up before proceeding to handle milking equipment or other setup tasks. You also can wash your hands during milking chores without having to leave the room. Near the wash sink you can install a hose faucet for washing down the floor and milk stand after chores. It is imperative to know approximately how much wastewater you will be creating and what to do with it so that you do not negatively impact nearby waterways, underground water sources, and the environment in general. Small licensed dairies are often

TIPS FOR UPGRADING AN EXISTING HOME DAIRY MILKING PARLOR

If you are milking cows or goats already and don't have the opportunity to build a new parlor, there are plenty of inexpensive fixes you can do to improve the conditions of a home dairy. To start with, look for ways to segregate the area from feed and animal housing areas. This might be as simple as hanging a vinyl strip curtain or building a partition wall. Don't ignore the ceiling; open rafters should be avoided. The parlor does not have to be closed off if you can locate it far enough away from animal areas and other spots that might be exposed to feed dust and other potential contaminants.

Once you have done what you can to close off the area, evaluate what you can do to make the surfaces in your milk parlor "cleanable." This might be as simple as putting on a couple of coats of glossy paint or hanging sheets of a waterproof material such as shower surround panels, dairy board, or fiberglass reinforced panels (FRP).

After that consider the floors. If you are milking goats or sheep, not only are they less likely than cows to relieve themselves in the parlor, but you can sweep up the dry pellets after milking. However, if you are milking bovines who are rather casual with when they decide to relieve themselves, proper manure cleanup is going to involve a decent amount of water, so you must provide a place for this dirty water to exit the parlor area, then be handled properly. (Wastewater management is an important topic, both for the cleanliness of your dairy and for the proper management of natural resources.)

For a home parlor with a wooden or dirt floor, and no option for pouring a concrete floor with drains, the minimum that should be done is to cover the floor with heavy rubber stall mats (designed for horse barns). If you cannot hose down the floor, you can periodically dust agricultural lime on the surface, then sweep (the FDA recognizes this as an approved method for bacteria control).

required to create official wastewater management plans—an onerous task, but one that serves an important purpose. Even if you are not to be licensed, you can be proud of managing your dairy's waste in a professional, earth-friendly fashion. See chapter 2 for more information regarding planning for dairy waste.

Milking Equipment

It might seem as if a milking machine allows you to collect milk in a much cleaner fashion than hand-milking, but this is true only if you religiously clean and maintain it; otherwise, the equipment, with its many parts, nooks, and connections, can contribute to extremely high bacteria counts and related problems.

A milking machine, when used properly, is very gentle and comfortable for the animal. In fact, it mimics the way that the animal's baby removes milk from its mom, through vacuum and rhythmic massage of the teat. I remember that when we got our first milking machine, I expected our goats—who, as all goats do, share their opinions quite readily—to put up a fuss before they got used to this new contraption. Instead, they acted as if it was nothing new, and some were even better behaved than they were when being milked by hand.

You can't properly use and maintain a milking machine without first understanding its components and how they work. So let's go over some of the parts

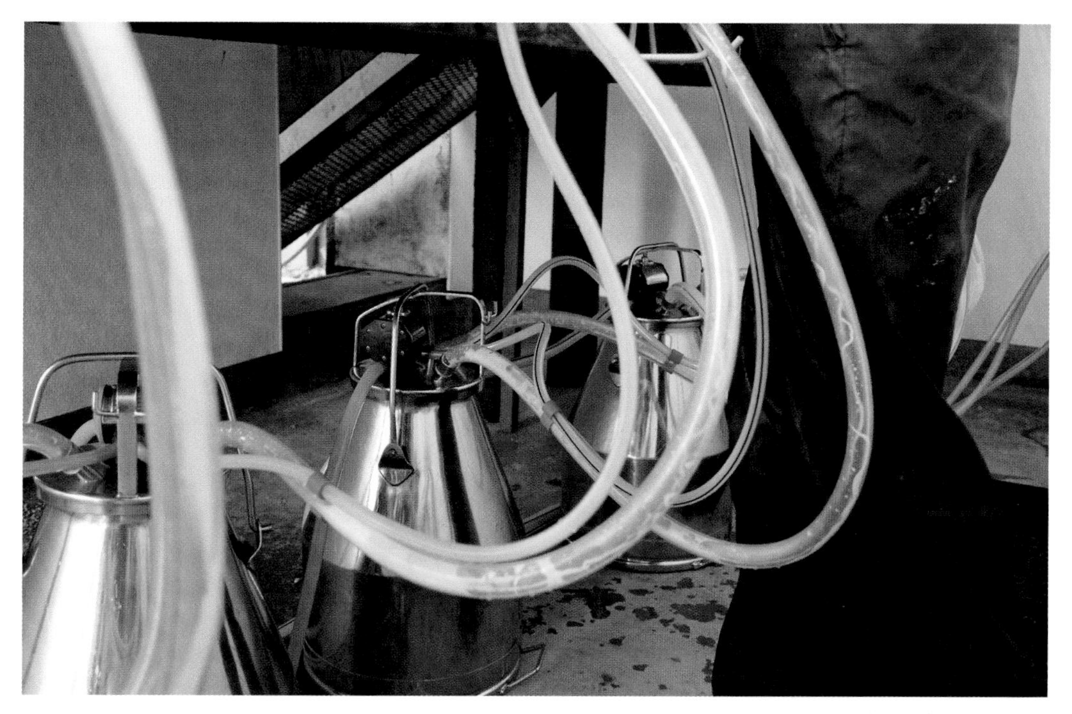

Here several portable buckets collect the milk from multiple animals being milked at one time on an elevated platform. The buckets are then poured through a filter and into a small bulk tank at Garden Variety Cheese, California.

and options of a typical portable bucket milking system with an eye toward making sure that it is working correctly, instead of contributing unseen problems. (I'll cover the details of cleaning the equipment in the next chapter.) We'll look at the vacuum pump, motor, and balance tank; the milk can and pulsator; the milking lines and vacuum lines; and the milking cluster—inflations, shells, and claws.

The Vacuum Pump

To understand a milking machine, let's start at the end farthest from the udder—the vacuum pump. This unit is made up of three main parts: the pump, a motor,

HAND-MILKING ON THE LICENSED DAIRY

According to the PMO, Grade A standards allow for animals to be hand-milked. If you choose to use this method all or part of the time, milking into a pail that is "hooded" is required by the PMO and highly advised even if you are not an inspected facility. The hood, a partial covering, helps prevent debris that might be on the animal's belly or flanks from falling into the milk pail.

If you are milking more than one animal, you should have a larger, fully covered pail on hand to pour each animal's milk into so it is more protected during the duration of milking. It is a good idea to place a stainless milk strainer with a disposable filter directly in this larger pail and pour the milk through it. This removes any debris quickly, so that it doesn't "marinate" in the nice, warm milk. If your milking procedure takes more than an hour, you may want to begin chilling the milk by setting the larger pail in a bucket of ice water during milking.

THE BEAUTY OF BACKUP AND SPARE PARTS

If your daily routine counts on the proper function of mechanical systems such as a milking machine and electricity to operate critical equipment, then by all means invest in backup motors, spare parts, and a small portable generator in case of malfunction, breakage, or power outages. I know many farmers who have found out the hard way the value of having such backups, often at the cost of a batch of milk, not to mention an incredible amount of work. Other producers might delay the purchase of a gas or diesel generator thinking that they need to purchase one that will run their entire farm. Instead of spending thousands, though, you can use a smaller unit to power critical systems as needed. For example, you can plug the vacuum pump into it during milking and then use it later to keep refrigerators from warming up. Spare inflations and small milking machine parts also should be kept on hand.

and a balance tank. The motor runs the vacuum pump, which in turn creates a vacuum inside the balance tank. A gauge will be located on the tank and should be monitored frequently to ensure that the proper vacuum level is being used. Vacuum level is usually maintained between 10 and 12 (measured in inches of mercury).

Portable milking machines vary in the number of animals that can be milked based on the size and power of this part of the equipment. A general rule of thumb is that you need 1 horsepower (HP) for every milking unit. For example, a 4 HP unit usually can milk four animals at a time. Keep this in mind when shopping. Also remember that a larger machine always can milk fewer animals, but a smaller machine cannot milk more. Given the option, an oversize motor is better than one that is undersize. You often can find fairly inexpensive milking machines for sale that either have a very small or even no balance tank. I suggest avoiding these, since they cannot provide as consistent a vacuum. In fact, the very best pipeline systems include a variable-speed vacuum pump, which adjusts the vacuum to maintain a consistent value. A balance tank also provides a safety buffer between the vacuum line and the pump—if milk or moisture enters the lines and is pulled into the vacuum pump, it can damage this most expensive part of the system.

A typical vacuum pump capable of milking four animals at a time. Alhtough pumps come in many sizes and configurations, a good setup includes balance tank, motor, pump, and pressure gauge.

Vacuum pumps are manufactured in two basic categories, those that use oil and those that do not. Although those that use oil are messier and need cleaning, in general they last longer. Depending on how it is set up, a vacuum hose or pipe will run from the pump and provide vacuum to two parts of the system—the milk can and the pulsator.

A system that starts out as "portable" can easily become a permanent piece of equipment. The vacuum pump and motor are not only noisy; since oil pumps can be messy, consider locating this part of the unit outside the milking parlor and running PVC (white plastic pipe designed for potable water supply) lines from the unit to wherever you need vacuum. As long as the lines are well sealed, there should be no loss of vacuum over most distances. You can wire an electrical switch into the unit easily and place it near where you are milking so you can turn the pump on and off without having to walk to where the unit itself is located.

Milk Pail and Pulsator

Next in the system are the milk pail, or milk can, and the pulsator. In many cases the pulsator—which provides vacuum pulsation (the intermittent on and off of the vacuum) to the inflations—is located on top of the milk can, but it can be a separate wall-mounted unit or can be a part of the milking hoses (which is the case with a newer type of claw that also provides pulsation). The milk can serves as a temporary "bulk tank" for the pooling of multiple animals' milk during the collection process. Although some states allow for the use of clear, food-grade plastic milk cans, I recommend using stainless steel if the milk will be used for human food. For our dairy we purchased several straight-sided Grade A milk cans and had one lid converted to work with our milking machine. (See the photograph of a modified milk can in chapter 9.)

Although the straight-sided cans are not primarily designed to withstand a vacuum within, they do it fairly well, but they will not last quite as long as the bell-shaped milk pails, which can better resist the inward pull of the vacuum. That being said, ours have been in steady use for six years without any adverse affects. The flexibility that we have in being able to milk directly into the container that holds the milk until it is processed makes this usage worthwhile for us. Both the milk can and the pulsator need to be supplied by the vacuum. So a vacuum hose usually runs to a connection on the lid of the milk can and a lid-mounted pulsator shares the vacuum supply to the can. If the pulsator is not located on the can, it will need its own vacuum supply line.

A couple of more words about lid-mount pulsators: I personally have had, and heard of, mostly problems with these units, even when rebuilt and maintained as advised. I know one cheesemaker who had her husband build a special little hammer with which to whack the unit (loosening the moving parts that tend to hang up easily inside). Instead of a custom hammer, consider a wall-mounted electric pulsator unit or in-line, individual pulsators such as made by NuPulse. Once we replaced our lid-mount version with this type, our milking lives became so much more trouble-free. Although the initial cost is higher, the lack of need for rebuilding and replacing quickly paid for itself. I have not used the in-line pulsator/claw combos, but I hear good reviews on their performance.

Pulsators can be adjusted for the rate of pulsation—that is, how many complete pulses occur per minute—and for the ratio between the squeeze and the release. Producers and manufacturers differ greatly on what they believe to be the best rate

WHAT ABOUT INEXPENSIVE, HAND-PUMPED MILKING "MACHINES"?

There are several setups available for purchase, as well as online instructions for building your own, that allow for the collection of milk through a hand-pumped vacuum and tube setup. (If you have ever been a lactating mother wanting to provide breast milk to your baby while you are absent, you will have used a similar device called a breast pump.) Although these systems do pull the milk effectively out of the teat, they are not designed for the long-term health of the teat and udder.

If you remember back to chapter 5 and the anatomy of the teat, picture the tissue of the teat and its blood supply. The hand-pumped systems provide a vacuum pressure that not only pulls the milk out of the teat but also keeps blood pooling at the end of the teat, rather than rhythmically releasing the pressure and allowing the circulation to flow. The original milking machines of the 1900s had similar problems, leading to teat damage and mastitis. A person can compensate for this to some degree by making sure that the teat gets regular breaks from the suction pressure, but I still advise that these units should be used only infrequently.

and ratio for milking different animals. I come from the school of a rate of sixty or fewer pulses per minute, even for goats, the teat time to completely refill, and a ratio of 60:40 or 50:50 (for example, if the pulse rate is one pulse per second and the ratio is 50:50, then the vacuum is open for ½ second and closed for ½ second). In conjunction with vacuum pressure, which I mentioned earlier, the rate and ratio may need to be adjusted over time, and no single setting will be ideal for every animal in your herd.

Milk Hoses, In-Line Filters, and Vacuum Hoses

Milking hoses and vacuum hoses run from the milk can and pulsator to the milking cluster. Milk hoses can be purchased in a couple of sizes; ⅝-inch lines usually will support the amount of milk coming from one or a couple of goats, sheep, or cows. The hoses come in plastic or silicone for Grade A milk. Although plastic hose is less expensive, I recommend always using silicone because it is flexible in even the coldest weather, will not stretch out at connections quickly—leading to hoses popping off during milking—and does not need to be replaced as frequently as plastic. If you use an iodine-based sanitizer, you must use plastic, not silicone, because iodine is corrosive to silicon. You can expect all your plastic hoses to become stained, but this is an aesthetic, not a quality, problem.

In place of pouring milk through a filter after milking, you can purchase in-line filters designed for portable bucket milking systems. These are basically small versions of the in-line filters used in pipeline systems. They are made up of a two-part plastic exterior, a rubber gasket/stopper, a stainless steel coil, and a disposable filter "sock." The sock goes over the coil and slides into the plastic shell. The coil keeps the sock from being sucked together by the vacuum from the bucket.

I love these in-line filters because they remove a step from my labor (lifting and pouring heavy milk cans) and also eliminate an opportunity for the milk to become contaminated while it is being poured. The filter shells currently available, however, have a silly design flaw: They allow for the tops to be overtightened to the point of cracking—easily! We went through several sets before figuring out what we were doing wrong. You should tighten the tops just to a gentle snugness. Even then, order a couple of extra sets of tops, just in case.

Vacuum pulsator hoses will be small, usually ¼ inch, and either black or clear vinyl or plastic. I recommend using clear sections in all or portions of the lines so you can see if moisture enters the lines—either from cracked or damaged inflations or from inadvertent entrance during washup. Not only can this help prevent damage to your pulsator, but it can tell you if there is a leak in one of the inflations—which can lead to high bacteria counts by basically making the inflation uncleanable.

The Milking Cluster

Now we get to the business end of the setup, the milking cluster, which is made up of the teat cups (shells and inflations) and a milk pooling section called "the claw." Not surprisingly, a cluster designed for cows looks different from one for goats or sheep. The typical cow cluster has a central claw (which resembles a bird or reptile's claw, hence the name) from which four short lines lead to each teat cup. Short vacuum lines run from the vacuum portion of the claw to each shell, where they provide pulsation to the soft rubber or silicone liner called the inflation. The inflation sits inside a rigid plastic or stainless steel exterior, called the shell. Let's talk a bit more about claws and their function; then we'll go over some teat cup options.

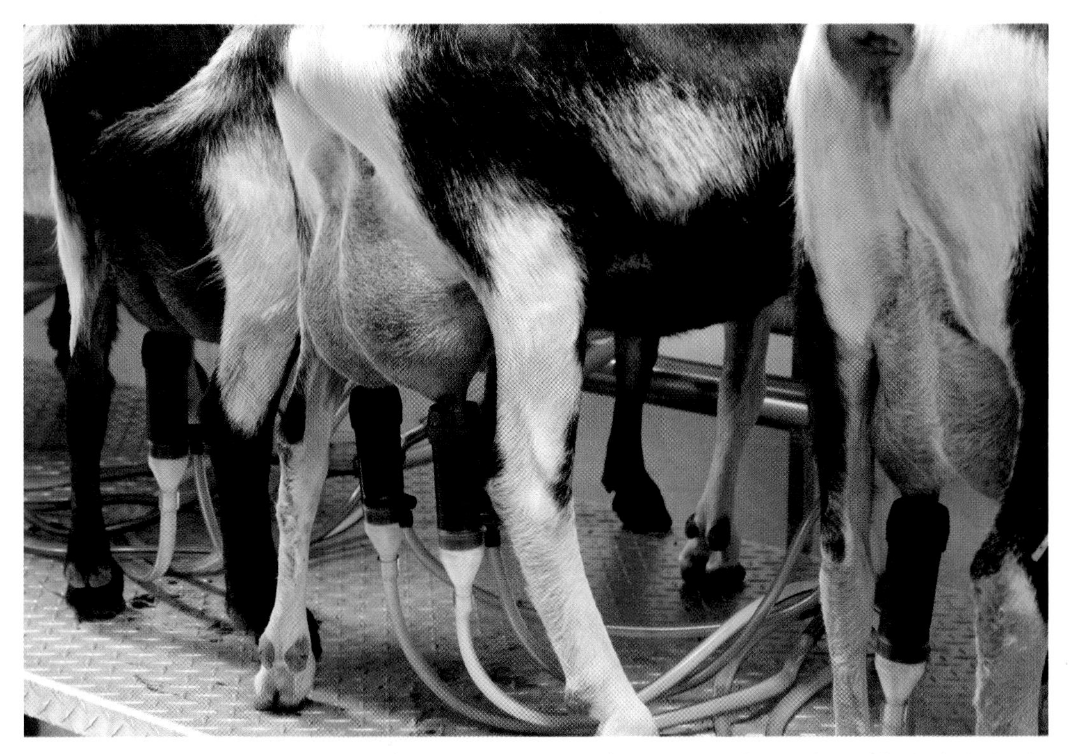

Goats being milked with all-in-one inflations (known as SP-6000 or IBA) at Rivers Edge Chevre, Oregon. Originally designed for cows, they work well on both cows and goats.

The claw's primary function is to provide a larger reservoir for the milk to pool in after it leaves the teat cup and before it enters the main milk hose. This prevents vacuum fluctuations caused by too much milk entering the line at one time. Claws have a tiny air vent that pulls outside air (from your super clean parlor, presumably) into the milking hoses. This allows the milk to pass through the hose without forming any blockages, which cause vacuum fluctuations that can allow milk to shoot back up into the teats, causing damaged teat orifices and risking mastitis through the spread of an infection from one animal to another.

Claws almost always include an "automatic" shutoff valve that will block the vacuum from the milk hose if vacuum is lost at the teat—if, for example, an animal kicks off one of its teat cups. This shutoff valve prevents, or limits, the suction of debris into the teat cup should it fall off. These valves must be opened when the cups are being attached to the teats to allow the cups to hold onto the animal. Some milking setups that work well for lower-output-volume animals, such as some goats and sheep, do not include a claw. This can work if the following is true: There is a vent in the line, usually at the base of the shell, that allows air to be drawn into the line; the milk hose line is of adequate size to accommodate the milk from all the teat cups, and the length of the line from teat cups to main hose is not too long; and the slope of the lines to the larger hose is downward. The SP-6000 inflation/teat cup (a similar product was made by IBA) has a built-in reservoir and all-in-one liner and shell and can be purchased vented (for use without or with a claw) or unvented for use with a claw.

Traditional teat cups are composed of a stainless steel or plastic shell into which a rubber or silicone inflation is placed. The inflation is followed by length of tubing, known as the short hose or short tube. The shell protects the inflation from wear and tear and provides a chamber in which the vacuum can pulsate to

TABLE 8.1. Milking Machine Parts: Cleaning and Maintenance Guide

Part	Life Span	Daily Cleaning	Periodic Cleaning/Maintenance
Lid Mounted Pulsator	Rebuild (replace rubber parts and gaskets) when needed; depends on cleaning and hours (can be 6 months to several years); keep a rebuild kit on hand!	Rinse, brush with detergent, inspect intake for signs of milk entering pulsator	Weekly disassembly from bucket lid or line to clean and inspect
Plastic Milk Hoses	4 months to a year	Lukewarm rinse, alkaline detergent wash, acid rinse	(If needed) Disassemble, clean with hose line brush
Silicone Milk Hoses	1–3 years	Same as plastic	Same as plastic
Rubber Inflations	1–3 months	Same as milk hoses	Same as milk hoses; use inflation brush
Silicone Inflations (including all-in-one types)	1 year	Same as milk hoses	Same as milk hoses; use inflation brush
Claws	Several years; inspect for cracks; replace internal rubber gaskets annually	Same as milk hoses; inspect air intake vent and clear any blockage if needed	Same as milk hoses; annually replace rubber gaskets if present

massage the teat. Many portable goat setups have plastic shells instead of stainless steel. These are less expensive and allow for the visual confirmation that the teat is situated properly within the cup, but they can be too lightweight, allowing the cup to move upward on the teat as the udder empties. This is okay unless the tissue near the udder enters the cup, where it can be damaged by the massaging of the vacuum. It also can allow sudden bits of air in from above. Liner slippage, sometimes called "slurping" thanks to the sound it makes, is a leading cause of vacuum fluctuations, which, as I mentioned before, can lead to teat damage and mastitis. So a "no slurping in the parlor" policy is a good idea!

The all-in-one shell and inflation type, such as the SP-6000 series, offers a great compromise of silicone inflation in a triangular shape and the weighted plastic shell. Some producers swear by these and claim reduced somatic cell counts. Although designed for cows, the smallest size works on most goats and sheep, even our farm's little Nigerian Dwarfs (but we modified them by installing little plastic elbows just below the inflation). If you use a cluster that does not include a claw and therefore not a shutoff valve, you should install in-line valves to turn the vacuum on and off when attaching and removing teat cups. Larger dairies using rubber inflations often simply bend these lines to turn the vacuum off. (Keep in mind that rubber inflations should be replaced as frequently as each month because they deteriorate more rapidly than any other material.) Check with your supplier to decide upon a replacement schedule based on both number of milkings and cleaning procedures.

Holding Pens, Platforms, and Headgates

Even the most docile animal probably will need to be secured during milking. If you are milking more than one animal, the setup will be more elaborate. The flow of animals coming and going from the milking area is also surprisingly important and often awkward unless properly designed. Trust me when I tell you that good traffic flow in the milking parlor is critical for maintaining the sanity of the small commercial dairy farmer!

Holding Pens

When considering the location and position of your milking platform or area, think about how efficiently you can move animals to and from the milking parlor. Here are some elements to consider:

- Distance from the main pen to the parlor
- Cleanliness of the area where animals might stand waiting to be milked
- Gates that are easy to move animals through—without having others escape or block the passage
- Pens that limit "return customers"—animals that try to return to the parlor for a second serving of grain

These does at Tempo Farm in Oregon provide milk for a local cheesemaker. Owner Lauren Acton, also a veterinarian, designed the parlor so that the animals stand at an angle, making good use of the space as well as providing easy access to their udders from the side. Note the use of inline NuPulse pulsator/claw units.

A step-up-style parlor can be used for cows, goats, or sheep, with a little initial training and encouragement of the animals.

- Holding areas that keep unmilked animals in one place while awaiting milking should have ample room to prevent bullying or stress.
- Comfort for you and the animal on the way from the holding or waiting area to the parlor; that is, shelter from severe elements or other opportunities for stress

Platforms

An elevated platform for the animals or a recessed standing area, or pit, for the people doing the milking can increase physical comfort and efficiency for the humans. If you're home-milking a single cow or just a few cows, a platform is not necessary. If you're milking goats and sheep at home, however, the use of a platform or stand, combined with a way to secure the animal's head in a stanchion or headgate, will be a great assist (more on those in a bit).

You can build many different configurations of platforms that will handle several animals at once. In fact, your design decisions are limited only by your ingenuity and access to space and materials. Here are some factors to consider when you are designing and building platforms:

- The number of animals will you be milking at the peak time of year
- Even if you can milk only one at a time, will having several standing ready speed up the process so that you don't have to make multiple trips to and from the holding area?
- Make sure animals of all age ranges can mount the stand or platform easily
- Space the animals so you have easy access for milking and so they do not jostle each other.

Headgate

A headgate, or stanchion, secures the animal's head and neck to keep it from moving too much during milking. For goats and sheep a variation on this is the sequential stall parlor in which the animal's entire body is held steady in a partially separated stall. Headgates for multiple animals usually secure them in one of two ways—"gang" and automatic. Gang refers to a single lever or line that is operated to close the entire group of stanchions at one time. Automatic headgates close individually once the animal's head is in place. Automatic headgates are also often "cascading," meaning that they close in a sequence that prevents animals from entering all but the next available gate. Most headgate systems allow the operator to pull a lever or switch that releases the entire group of animals at one time.

Most dairy farmers feed a bit of grain to the animal during milking. In that case, a pail or trough is included to hold the grain. One piece of advice: If you do feed grain during milking, feed only as much as you decide is the right amount. Goats, in particular, are good at "training" their people to think that only more grain will keep them content and quiet until milking is finished. Many people have told me apologetically that they can't milk fast enough and have to keep giving their goats grain, hay pellets, or other treats to keep them from misbehaving. A well-trained

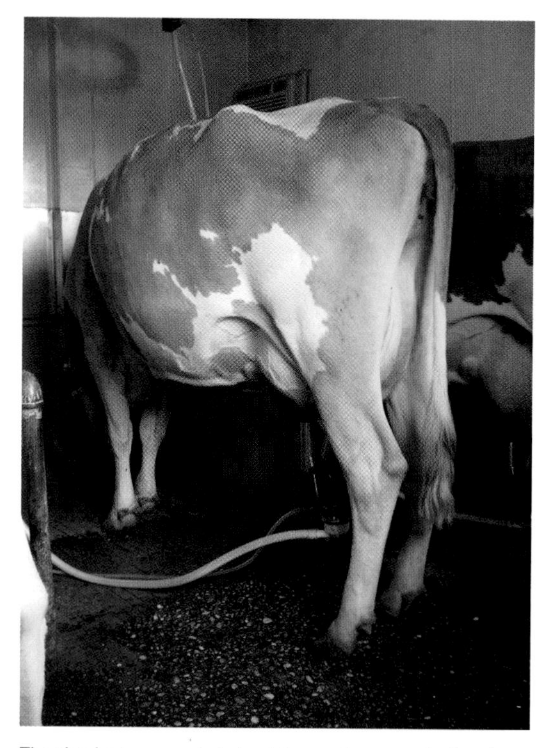

The simplest approach, but not the most ergonomic, is a tie stall barn. Here Guernsey cows are milked three at a time at Sweet Home Farm in Alabama.

goat, sheep, or cow will stand for extended periods of time with no feed present.

Here are some factors to consider when you are choosing a milking stanchion:

- The number of animals you will be milking at the peak time of year
- The speed and efficiency of securing and releasing more than one animal at a time—"cascading" headgates and stalls secure the animals in an orderly fashion—preventing traffic jams
- The height of the headgate and the ease with which the animal can get its head in and out of the stanchion
- The spacing between animals so that they do not steal grain or antagonize their neighbors

Monitoring Milk Production

There are many benefits to monitoring and documenting the milk production of your animals, even if you have a herd of only one. From the most intensive, data-rich program, called Dairy Herd Improvement (DHI), to a simple milk scale and notebook, there are many degrees of milk-production recording. The most obvious information collected is the volume of milk that each animal produces, not only day to day, but seasonally and over her career. Knowing milk-production volume over this spectrum can help you devise feeding programs, anticipate volume for sales and product production, and choose good genetics. There is probably no sensible dairy farmer out there who wouldn't trade 10 mediocre animals for fewer that can produce the same amount of milk.

Beyond volume the data provided through a formal milk-production program includes things such as butterfat, protein, and lactose production for each individual milk animal, further helping the farmer choose animals and tailor feeding programs. For the farmer producing milk for cheesemaking, the number goals might be different from those for a fluid milk producer. DHI testing includes analyzing the milk for somatic cells, or a somatic cell count (SCC). By closely monitoring SCC, the farmer can develop standards for her own herd—and for individual species and breeds—that will allow her to assess milk quality and udder health over time.

For the true data geek, DHI paperwork is a mother lode of management information—some of which, even after 10 years on the program, I don't fully comprehend. Still, I enjoy poring over the data, using it to make breeding decisions, and I frequently learn something new that helps us better manage our herd. Information about peak production days; production of fat, protein, and somatic cells by age and lactation group; and reproductive information—who they were bred to and how many babies they had—are a few of the examples of what DHI data can provide. In addition, the simple fact that this information is "official" lends an air of legitimacy to your claims regarding milk volume. When selling

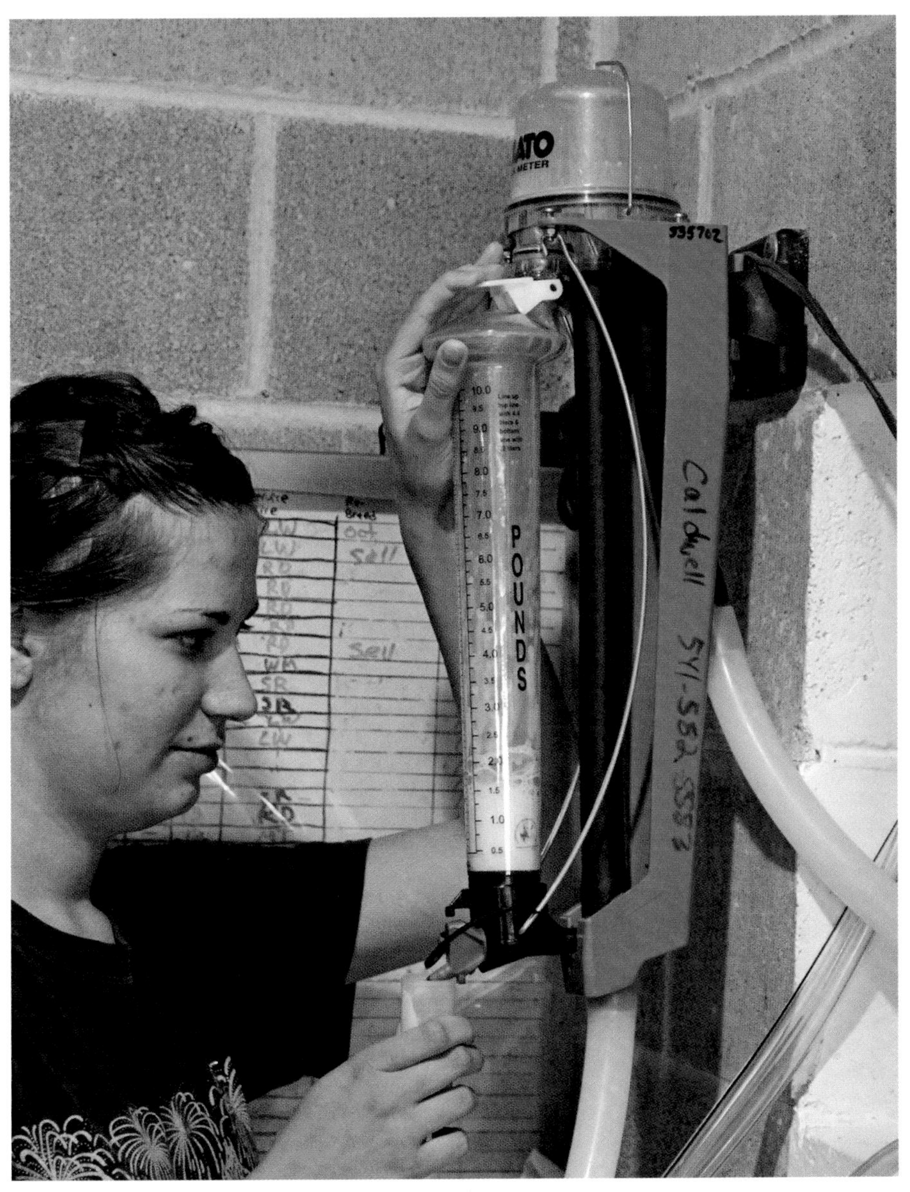

A milk meter, such as the Waikato meter pictured, allows for milking to be done by machine and still provides an accurate measurement of milk production. The milk tester draws a sample after the animal is done being milked.

offspring or adults, the value added to the animal's sale—and hence your pocketbook—is worth the time and trouble.

Milk volume also can be measured using a hanging scale and milk pail. When using a milking machine, special milk meters can be used. For official records all scale types must be calibrated for accuracy and certified by an approved milk testing association. Most hanging scales can be set to read zero when the milk pail is set in place, making it easier to read only the milk weight. Milk meters usually are designed to collect a calibrated sampling of the milk as it flows through the meter. This sample accumulates in a vial that correlates with a volume reading and a milk sample, for testing by a laboratory, and can be drawn from the vial once the animal is done being milked.

· 9 ·

The Milk House and Bottling Room

In most conventional dairies today, milk is picked up by a milk tanker truck and taken to a larger milk bottling or product plant. Until that time the milk is stored in a cooling unit called a bulk tank, which is located in a room called the milk house. Even on smaller dairies where the milk will be bottled on site or made into cheese, a bulk tank usually is located in a milk house and stores milk from several milkings. This same room is where the milking equipment is cleaned, sanitized, and stored (unless a clean-in-place pipeline system is used). If you are milking for home use only, you may be able to place the washing area in the parlor (if it is well sealed from dust and outside contaminants) and use one room for milk storage and bottling. But the best choice, even for a tiny farm, is a washroom separate from the parlor and a bottling and storage room separate from all other rooms.

Let's talk about some construction tips, equipment, and supplies for keeping your milking equipment sparkling clean, and some options for milk chilling, storage, and bottling. At the end of the chapter, I'll briefly cover some of the other rooms that a well-functioning small dairy may find useful.

Milk House Construction Tips and Considerations

I like to tell people that construction in a dairy can be summarized by a simple tenet—each room that represents a step forward in milk's journey to the consumer must be proportionately better constructed, easier to clean, better illuminated, and more protected from outside contamination. If you keep this in mind, it isn't too hard to look at the space, compare it to the milking parlor, and bring it up to a higher standard.

The door between the milking area and the milk house should be self-closing and have a snug fit to prevent dust from entering the room. Windows should not open to any animal housing area. If you already have windows in this room and need the ventilation, consider using a large filter in the open space. Filters designed for air handlers (part of a whole-house heat pump system) work well.

Construction in a dairy can be summarized by a simple tenet—each room that represents a step forward in milk's journey to the consumer must be proportionately better constructed, easier to clean, better illuminated, and more protected from outside contamination.

It can be a revelation to see just how much dust can accumulate, another one of those things that we don't notice if we cannot see it!

Be sure to provide the best-quality ceiling exhaust fan, designed for moist areas such as showers, that you can afford. If you are really clever (we weren't), you will include a timer switch for this fan so you can turn it on and have it run for a period of time after you are done cleaning the milking equipment. Remember that a dry area is not likely to mildew or encourage the growth of unwanted bacteria. But a moist area will not only encourage these things; it will mean a lot more cleaning work for you.

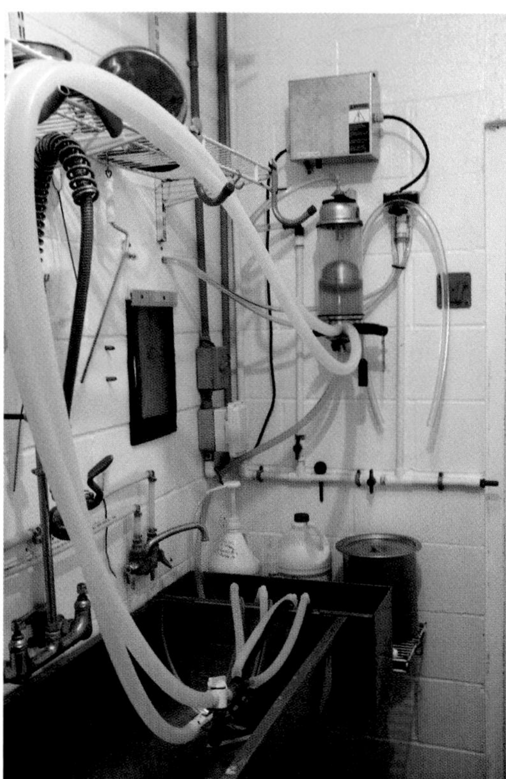

If you are lucky enough to have a floor drain in this room (and if you are a licensed dairy, you must have one), be sure to locate it where it will not be covered by any equipment, such as a bulk tank or freezer.

Even if you are not going to be licensed, a handwashing sink in this room is a good idea. Sometimes it is hard to see the point of so many sinks in a dairy, but they indicate the best intentions in steps of sanitation that all add up to clean, safe milk. And fortunately, you can buy some inexpensive little handwashing sinks that do the job just fine.

A so-called bucket washer (designed to wash the hoses from a bucket milker) can help make the cleaning of equipment efficient and simple. The milking clusters hang in a small sink in which the different solutions for each stage of washing are mixed. The washer unit is connected to the milking machine vacuum supply, which pulls and pushes these solutions through the lines.

Keeping Things Clean

The most important ingredient for proper cleaning of all milking equipment is a good source of very hot water. If you don't already have a hot-water tank, I recommend installing an "on-demand" or "tankless" water heater. These can be purchased in models that run on electricity or propane or natural gas. Not only will these units supply you with plenty of hot water, but they will help you save on your electricity bill as well.

If your facility will be licensed, the style of sink you install, whether that be a single, double, or triple basin, will be dictated by local regulators. If you are not licensed, a large, single sink works quite well. If for some reason you don't have a separate handwashing sink, by all means choose a large two-basin sink and dedicate one side to your handwashing and the dirtiest stage of washing the milking equipment. If you are routinely washing the equipment by hand, the basins will need to be large enough to soak all the lines and hoses as well as accommodate the milk can. I adore having a commercial restaurant-grade sprayer to use when washing milk cans. Even if you cannot afford one right away, put it on your wish list, and leave room for a future installation.

CLEANING AND SANITIZING PROCEDURES: THINGS TO REMEMBER

- You cannot sanitize something that is not clean; therefore, proper cleaning is as important as proper sanitizing.
- Proper dilution of detergents and sanitizers must be coupled with proper water temperature, scrubbing, and length of time cleaned to be effective.
- Sanitizers not diluted to the proper concentration will be ineffective if weak and will leave a residue (a chemical hazard) if too strong, as well as damage some equipment.

You may want to add a system that is specifically designed to aid in the proper and practical cleaning of your equipment. Lines can be cleaned effectively manually after each milking by using brushes, but this will erode the interior surfaces much more quickly than a vacuum washing system, and it also will be time consuming for you. The bucket washer (so called not because it washes your buckets—it doesn't—but because it is meant for use with bucket milker systems) consists of a unit that mounts to the wall onto which you attach your milking lines. A small sink or pail is filled with the washing solutions and the inflations are immersed in this pail. A line runs from the wall unit to your vacuum source. When the vacuum is turned on, the unit pulls and pushes the washing fluids through the lines. Between each cycle (pre-sanitize, rinse, wash, or acid rinse), you will turn off the suction and change the solution in the small sink or pail. The buckets and lids are scrubbed by hand. At a larger dairy a clean-in-place (CIP) system does basically the same thing, but forces the solutions through the built-in-place milk lines.

Step-by-Step Dairy Equipment Cleaning

1. Before milking, sanitize equipment with a no-rinse, food-contact-surface-approved sanitizer. Verify the proper dilution strength using a test strip.
2. After milking, rinse all equipment with 100°F (38°C) water to remove visible milk residue.
3. Next, using very hot water (the right temperature will vary based on the chemicals you are using, water pH, and water mineral

Allowing milking equipment to dry between uses is an important step in milk-collection hygiene because bacteria require moisture to multiply.

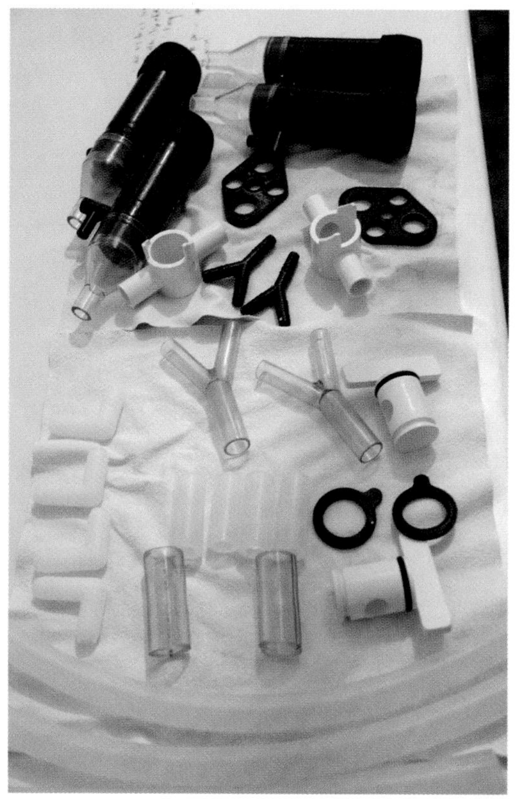

Complete disassembly and cleaning of milking equipment should be done periodically.

content—hard or soft) and detergent, wash equipment thoroughly. If you are hand-washing, soak everything in the hot water and detergent for 5 to 10 minutes, then scrub. If you are using an automatic washer, run the cycle for 5 to 8 minutes. Don't let water temperature drop below 120°F (49°C) or proteins and fats may be redeposited.

4. Rinse with 100°F water OR rinse with properly diluted acid sanitizer rinse.
5. If you don't use an acid rinse daily, perform a weekly acid wash (see Table 9.1) to remove mineral deposits.
6. Hang the equipment to dry.

Chilling and Storing Milk

Chilling milk—and keeping it cold—is an essential step for milk quality and safety. In fact, it is probably the one step that is the most poorly addressed in most home and small dairies. For the small processor chilling milk presents some special challenges, since most equipment is designed to handle larger volumes of milk. Luckily, there are some companies now designing and manufacturing small equipment for such situations. Although I am covering this equipment under the milk house heading, in many situations it will make more sense to store cold milk in the bottling room. Let's talk about some options for making sure that this critical step in milk production is done properly.

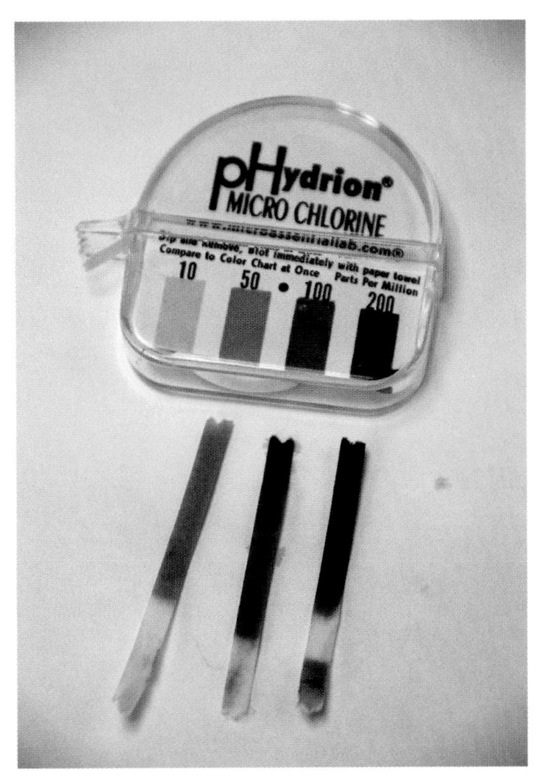

Inexpensive chemical titration test strips will help you properly measure and check the concentration of chemicals. The test strip on the right shows a concentration more than twice as strong as it needs to be (based on the commonly recommended measurement of ½ oz [1 Tbsp or 15 ml] of household bleach) per gallon of water.

Bulk Tank

A bulk tank stores the milk from several milkings, or at least it is capable of doing that. Typically, milk flows directly from the milking parlor, through filters, and into the bulk tank, where agitators stir the warm milk until it is chilled to the proper temperature, then automatically stirred at intervals. Chilling is usually accomplished with a built-in compressor or a remote cold-water source. Most conventional tanks require a 220- to 240-volt power outlet (like the ones used for an electric clothes dryer), but some new units come in small sizes that require only a regular 120 volt outlet. The unit

TABLE 9.1 Cleaning and Sanitizing Guide for Dairy Equipment

Name	Common Chemical	Purpose	Water Temp	Cycle Run Time
Detergent	Potassium or sodium hydroxide (lye) (pH 12)	Dissolving of fat	120–160°F (dependent upon detergent and water)	Wash cycle, 8–10 minutes
Chlorine Added to Wash Cycle	Chlorine (50–100 ppm)	Dissolving of protein	Same	Part of above cycle
Acid Rinse/ Sanitizer	Phosphoric acid, phosphoric-sulfuric blend, phosphoric-citric acid blend, peracetic (peroxyacetic) acid (pH 2.0)	Prevention of calcium deposits (milkstone); neutralizing of alkaline cleaners, acidic residue, for between milkings	70–110°F	Acid rinse cycle, 2–5 minutes
Acid Wash	Same as for rinse, but with the addition of surfactants for additional mineral removal power (pH 2.0)	Used periodically after alkaline cleaner cycle	110–120°F	Acid wash, 5 minutes
No-Rinse Sanitizing	Chlorine (50–100ppm), iodine 12.5–25 ppm), peracetic (peroxyacetic) acid (82–197 ppm)	Presanitizing of equipment and milk hoses	100°F or cooler	Presanitizing, 30–60 seconds immediately before use (no rinse needed)

This small bulk tank is typical of the old, refurbished milk chilling and storage tanks used on small dairies.

should be all stainless steel, with no seams or connections, since those are difficult to clean. It likely will be well insulated.

Normally, the first milking to enter the tank will need to have a volume of about 10 percent of the tank's total capacity for the agitators to make contact with the milk. Some producers are able to bypass this by hand-agitating the milk, but this is an awkward step that you should avoid if possible.

If a tank does not provide its own water cooling, you'll need a "remote chiller" as part of your system. Commercial remote chillers come in a range of prices and power consumption. You can improvise your own remote chiller using a chest freezer and aquarium or water-feature submersible pump. The chest freezer should be prepared by sealing all the interior seams with a high-grade silicone caulking. You can drill two holes in the lid (if you are drilling from the side, be sure to verify that no electrical or compressor lines run through the walls!) to provide an outlet and a return for the chilling solution.

An effective chilling solution can be made using water and food- or cosmetic-grade propylene glycol (purchased most cost effectively online in 5-gallon drums). The propylene glycol keeps the water from freezing and chills the solution more rapidly than plain water would because it lowers the freezing temperature. (I have heard of rubbing alcohol being used as an antifreeze, but in an open chest such as described here, the alcohol would evaporate and need to be replenished fairly frequently). The solution will need to be diluted so it flows through the pump and lines. This may be at a rate of about 50:50, but observe how thick it is and monitor the pump's effectiveness. Chest freezers cool in cycles as well, so sometimes the thickness of the solution will vary.

Immersion Cooling

If your volume of collected milk does not warrant the use of a bulk tank, or if you are not storing milk for very long, you can use the immersion method of chilling, which predates the use of bulk tanks, and although it's more basic, it can be just as effective. There are many approaches to this method, from using chilled water from a spring to setting pails in a freezing slurry of water and propylene glycol. You'll need to stir any milk that you chill in this fashion to be sure it chills evenly and thoroughly in the same amount of time as required for licensed dairies (to under 40°F [4°C] within 2 hours of milking). Even if you are not licensed, milk quality is best preserved by rapid chilling.

Once the milk reaches the goal temperature, you can move it to a storage refrigerator or bottle and store it. You will need to allow adequate time between uses for the chilling solution to reach a cold enough temperature for adequate chilling by the next milking. By the same token, you may need to increase the total volume of chilling solution as the amount of milk increases, or chilling times will grow longer. Since milk leaves the animal at around 100°F (38°C), it takes quite a bit of cold solution to decrease that temperature to the goal.

A modified milk can (the stainless steel tubes were added to the lid) serves as a bulk tank of sorts when immersed in a bath of food-grade propylene glycol. The milk flows through in-line filters and directly into the can.

Monitoring Temperature

Whatever method you choose to chill the milk, monitoring and recording the temperature is an important step toward both ensuring and proving that you are caring properly for your raw milk. Chart recorders are the industry norm for use on ready-made bulk tanks. They also can be adapted to immersion chilling systems. In place of a chart recorder, and where approved, a digital temperature logger can be used. These units include a temperature probe that you place in the milk, then begin to record the temperature. The unit keeps track of many "sessions," and when its digital memory is full, you can download the data to a computer and keep a record of milk chilling. Although it's inexpensive (compared to a chart recorder), there are drawbacks to this method, such as keeping the temperature probe clean and sanitized, needing to replace batteries, and needing to download and maintain digital records. For home produc-

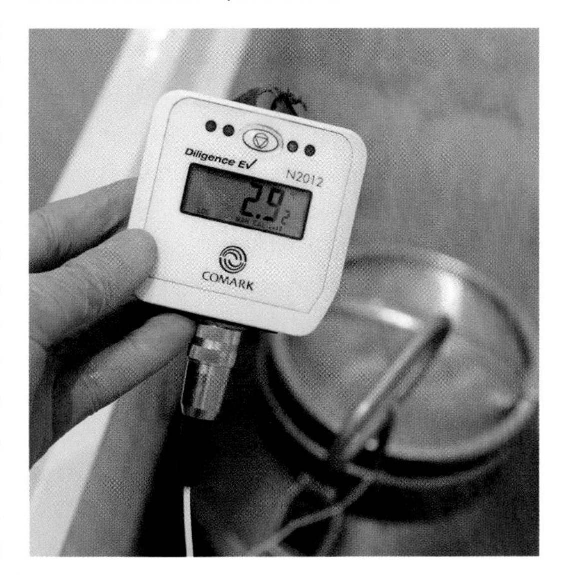

A digital temperature logger records temperatures at set intervals. When the instrument's memory is full, the data can be downloaded and saved to a computer or other memory-storage device.

ers you can simply monitor the temperature with a handheld, hanging, or floating thermometer. I don't recommend using a glass thermometer because the risk for breakage is high and glass shards in milk pose a serious risk to your life and health should the break go unnoticed, not to mention having to throw away milk if there is a break.

Bottling Room Construction Tips and Considerations

By now you probably will anticipate my advice for constructing a proper milk bottling room: one that is better lit, more easily cleaned, and more secure from outside contaminants than any other room in your small dairy. You also should start thinking about the traffic flow of people and also product, from when it enters the room as chilled milk to when it leaves as a product ready for the consumer. For example, if you carry milk cans from the milk house or parlor, can you ensure that feet and cans are clean and sanitary before they enter the bottling room? This may be as simple as having a small changing room or footbath and wiping the milk cans with a sanitized cloth before setting them into the bottling room and using a separate door by which bottled milk leaves the room.

It is useful to draw out your floor plan, using colored pencils to mark the traffic flow for both people and product, using one color for those coming in from the dairy side and another for those leaving with finished, consumer-ready product. People have been sickened by touching the outside of a milk jug that was contaminated, then handling food without washing their hands.

Ideally this room should also include a floor drain located so that the room can be thoroughly cleaned with all washwater running easily to the drain. As with the milk house, the drain should not be located under equipment or fixtures. Drains can harbor some of the most concerning contaminants, such as listeria, so be sure to keep them clean. The cleaning of floors and drains should never occur when milk bottling is occurring or when packaged milk is sitting unprotected in the room.

As is true with the milk house, having good air ventilation is important, and because milk will be processed and packaged in this room, even more attention should be paid to keeping humidity and condensation at bay. A properly sized exhaust fan is critical to removing humidity that will occur during cleaning. When deciding where to place the exhaust fan, keep in mind that condensation may occur in the air outlet and drip down from the fan, so don't place it where that moisture could drip and contaminate a surface, milk bottles, and so on. This type of cross contamination can occur without being noticed and can lead to serious problems for some susceptible consumer.

Also keep in mind that an exhaust fan will pull air into the room from any opening, whether that is a window, around a door, or another opening. You must make sure that any air entering the room is not bringing with it dust and other contaminants. If you already have a fan installed, a simple way to observe its effectiveness is to light an incense stick and walk around the room, placing the smoking incense into every nook and cranny and see where air is being pulled into the space.

Pasteurizer

For those who wish to sell low temperature pasteurized milk, you'll need a small legal pasteurizer. To be approved by your inspector, the unit either must come stamped with the 3A approval rating or must be physically inspected and approved by your individual inspector (or regulatory agency). Large-scale dairy processors usually are required to use only 3A-approved equipment, but smaller processors often can receive individual approval. The 3A sanitary standards define issues such as the quality of stainless steel and the way in which fittings are finished. (For more on 3A standards see the bibliography.) Before purchasing any equipment that does not carry the 3A rating, be sure to make the sale contingent on your inspector's approval. Remember that the term "pasteurized" can be used only when an approved, legal pasteurizer is in use in an inspected facility and the process is performed by someone trained and licensed to do so. Home pasteurizers are fine for personal use but cannot be used for milk produced commercially; in other words, available for sale.

Typically, the smallest commercial unit will hold about 15 gallons of milk. Most of these units are quite expensive, but MicroDairy Designs has a model that is reasonably affordable because it does not include a drain and therefore does not need an expensive component called a leak detection valve. This does mean, though, that the milk must be pumped out of the unit after cooling, which does require other equipment and approval by inspectors, but it can still be a cost-effective solution. In addition to the pasteurizing unit itself, you will need a chart recorder to document proper temperature and holding time and a water chilling unit or a cold-water source to chill the milk in the unit once pasteurization is complete. You also will need to obtain a pasteurizer's license (the pasteurizer's exam usually is taken on-site after the completion of a class).

Bottling Equipment

For the small home producer, bottling equipment probably will consist of a stainless steel filter holder and a glass milk jar; this may even be the equipment of choice for those producing milk for others under various states' regulations. In some states, however, even the smallest producer must use a machine to fill and cap milk containers. Even if you live in a state in which this is not required, it is worth looking into it as a step toward more professional production. The advantage of mechanical filling is not only that it reduces the milk's exposure to handling but that it also keeps milk from dirtying the outside of the containers. Mechanical capping helps reduce the possibility of contaminating your lids through handling. I believe that both of these steps can be accomplished manually, however, in an equally safe fashion with the right amount of vigilance.

The Eco-Flex packaging system from MicroDairy Designs has made small batch, economical bottling possible for many producers. *Photo courtesy MicroDairy Designs*

Thanks to the ingenuity of Frank Kipe of MicroDairy Designs, small producers can purchase an affordable and efficient system that meets most states' requirements for mechanical filling and capping. His system, called the Eco-Flex packaging system, can accommodate several sizes of milk containers and even fill yogurt cups. A vacuum pen is used to place caps on containers (see appendix A for contact information). As of this writing the packager is being redesigned to include an automated cap dispenser to satisfy the regulations in states in which using a vacuum pen to place caps on bottles is not acceptable.

Many producers use refurbished bottlers that are many decades old. These typically work well but often require two people to operate them properly. Sharilyn Reyna, profiled later in this chapter, and Alethea Swift, profiled in chapter 10, both use such equipment. Alethea, however, will be transitioning to an Eco-Flex packager, as her volume is better suited to a machine that one person can operate. A photograph of bottles being filled using a refurbished, antique bottler appears in chapter 10.

Milk Containers

Whether you hand-fill or not, choosing bottles and containers for your milk is a critical, and often personal, step. For many, glass is the most appealing choice, but for others, plastic makes more sense. Let's go over some of the pros and cons for each. We'll also cover cleaning glass bottles and the importance of adequate cold storage of bottled milk.

Glass

Glass takes the prize in the nostalgia category for milk containers, and it usually will command a higher price if you are marketing and selling your milk. Commercial producers will find very few options for purchasing new glass bottles. Sharilyn Reyna (see sidebar "Sharilyn Reyna, Licensed Goat Dairy, Oregon") says hers come from a company in eastern Canada that requires up-front payment. Reyna, who requires a $3 deposit on each half gallon she sells, says that typically only 50 percent of the bottles are returned. For the home producer glass is a reusable, readily available choice—with many people using widemouthed half-gallon jars.

In addition to limited availability commercial glass milk bottles bring some other issues as well. First, there is a risk of breakage and with that the possibility of introducing a physical hazard into your process. Reusing glass bottles and jars also requires a decent amount of energy—both personal and utilitywise—to clean and sanitize them properly. Glass also can cause nutrient loss and flavor changes if the bottles are exposed to light. Some small-scale producers require that their customers "buy" and own their own containers—returning them clean and ready for refilling. This is convenient and makes sense on one level, but keep in mind that if the container is not as sanitary as it should be, it is likely your milk will be blamed first, not your customer's washing techniques.

Plastic

Plastic might have a less than glamorous and wholesome reputation, but it does have some advantages over glass. It is inexpensive to purchase, easy to store, and nonbreakable and, in some cases, can protect the milk from light damage. Today's plastic milk jugs are free from known chemicals that could leach into the product. If you decide to go with plastic, you might consider including some educational labeling or offer your customers opportunities to help them understand and appreciate your decision.

Labeling

If people outside your immediate family are consuming your milk, it's wise to consider labeling the containers, even if you are not required to do so. In states that allow for the legal sales of raw milk, you no doubt will be informed of the labeling requirements. Most labeling requirements include warnings about the dangers the FDA and most state departments of agriculture and health believe are inherent in raw milk. For example, Washington State requires the following:

> WARNING: This product has not been pasteurized and may contain harmful bacteria. Pregnant women, children, the elderly, and persons with lowered resistance to disease have the highest risk of harm from use of this product.

SMALL DAIRY PROFILE: SHARILYN REYNA, LICENSED GOAT DAIRY, OREGON

Crowds form when Shari or her long-time partner, Fred, show up at the health food grocery store. No, they aren't movie stars—customers are there to get their hands on the real celebrity, glass bottles filled with Fern's Edge Dairy's raw goat's milk. The crisp, clean milk is so popular that some stores have resorted to posting signs that limit customers to a few bottles each. Although it is legal to sell raw goat's milk at the retail level in Oregon, Fern's Edge Dairy is the only farm providing this highly sought-after product.

Reyna's path to raw-milk royalty began when she was a small child and her allergies to cow's milk brought a dairy goat into her family's life. Later, in 1974, when she was a graduate student working on her PhD in archaeology in Oregon, Sharilyn added a few milk goats to her small homestead outside the lush college town of Eugene. To the young mother, a self-described hippie, goats helped complete the circle of self-sufficiency and connection to the earth that she sought for her life.

When I first visited Fern's Edge in 2006, the newly built creamery was making amazing artisan goat cheeses from the milk of Shari's goats. The herd had grown from the small home-milker group she started with in the '70s to around one hundred Alpine lovelies. Although the dairy officially was just getting going, Shari had been milking her goats year-round for several decades. The license simply made it official. Although cheese was their first product, the antique milk bottler that I saw during my tour was sitting ready in a room dedicated to the bottling of raw milk.

Currently the farm bottles about 100 gallons each week, with the rest of the milk going into the production of pasteurized and raw-milk cheeses, for which the farm is widely acclaimed far beyond the region in which the milk is sold. Fern's Edge milk is delivered weekly to stores located throughout the region. Shari, whom I have never seen without her Blue Tooth headset perched behind her ear, chatted on the phone with me about the farm's raw milk while she made the hour-plus drive to several of the markets on her route. She said their production is a constant balance. "If we make too much cheese, then we don't have enough fluid milk; if we sell too much milk, then we can't make enough cheese."

Reyna treats the process of raw-milk production no differently from that of milk that's headed for pasteurization for some of the cheeses made at Fern's Edge. In other words, she would put the same care into collecting milk destined for heat treatment. "Why would I not want all of it to be the best?" she chided me. Her care extends to all levels—milk for delivery is bottled in the freshly washed and sanitized glass bottles and stacked in milk crates that have also just gone through the Hobart hot water sanitizer (a 2010 pasteurized milk outbreak from a larger cooperative dairy of salmonella-related illness was traced back to contamination on the milk crates). Her partner Fred, a chemist, does on-farm testing of their milk, and Shari checks milk bottle temperatures upon delivery utilizing a handheld infrared temperature reading device.

Her herd of goats is well known in its own right—both for milk production and as show animals. Shari, whose activist heart still beats strong, is a dynamic voice in the world of cheese, raw milk, and, of course, her beloved goats.

To learn more about Shari, her goats, and Fern's Edge Dairy's award-winning cheese and prized raw milk, visit https://www.facebook.com/FernsEdgeDairy.

Sharilyn Reyna holding a bottle of Fern's Edge raw goat's milk.

California adds a bit to this with their required label:

> Raw (unpasteurized) milk and raw milk dairy products may contain disease-causing micro-organisms. Persons at highest risk of disease from these organisms include newborns and infants; the elderly; pregnant women; those taking corticosteroids, antibiotics or antacids; and those having chronic illnesses or other conditions that weaken their immunity.

Washing and Sanitizing

Depending on how you're bottling your milk, whether with a mechanical system or by hand, and whether you will be washing and reusing glass containers or not, you will need washing and sanitizing equipment to match those needs. For glass containers a commercial dishwasher can be very effective. Washing bottles shortly before use will ensure they are well sanitized by the heat.

When hand-washing, first rinse your bottles in lukewarm (95 to 100°F [35–38°C]) water. Then soak and scrub them in a hot, chlorinated alkaline detergent (see table 9.1 for more information on cleaning and sanitizing products), and scrub them thoroughly with a bottle brush. Rinse them with clean water. Just prior to bottling use, dip them in an approved food-surface, no-rinse sanitizer solution. A three-compartment sink is ideal for completing these three steps in the most effective way possible. As with the milk house, make sure you have ample hot water available to complete all your cleaning tasks. Lids also should be sanitized and air dried before use.

Once the milk is bottled and the containers are clean, sanitized, filled, and labeled (this must all occur without the milk temperature's increasing above refrigeration temperatures), the milk should be moved to a refrigerator and kept cold until it is consumed. If you have a farm store or honor stand, the milk can be moved into a refrigerated unit in that space. Customers should not have access to your bottling room or any other part of the dairy!

Other Rooms

Licensed producers and those wishing to have a well-functioning facility will likely want to have easy access to several other rooms or spaces. Not the least of these is a toilet facility with a handwashing sink. A place to cleanly store and protect any packaging materials is also essential. This could be as simple as plastic bins into which milk jugs can be placed immediately after arrival to keep the containers protected from dust from the dairy and possible contaminants from chemicals and other contaminants. If you will be performing any lab tests, which I highly recommend and will cover in the next chapter, a space away from milk storage or processing will be needed. Records and documents regarding your business and food safety program should be kept in a dry, well-organized space as well.

· 10 ·

Using and Understanding Laboratory Milk Tests

A few years ago I learned that with a few inexpensive and simple-to-use products we could perform our own lab tests—right here at our farm. Not only would these tests be affordable, but we would have the results almost immediately. Even after I learned how to do these tests, it still took us awhile to order the supplies and make testing a regular part of our production. Now I am not sure why it took so long. The knowledge it has given us of the effectiveness of our procedures has been priceless. So I hope after reading this chapter you will jump right in, order your supplies, begin testing, and utilize the information that these tests will provide. Power to the farmers!

First, we'll go over some of the tests required and/or performed by many states both for animals and for milk. We'll talk about how to interpret the results of these tests—exactly what the numbers mean—and what less than stellar results might mean and how to improve them. Then I'll talk about what tests you can do yourself on your farm, the supplies you will need, and how to read the results. Even if these tests are performed in an uncertified lab (your farm) and read by an untrained technician (you), they still can be powerful tools in the production of high-quality milk.

Stating the Standards—Tests Required by Regulators

The laws surrounding the legal production of raw milk, by both licensed and unlicensed producers, not only vary tremendously from state to state but also are constantly under scrutiny and subject to revisions and changes by each state's lawmakers. Often a high-profile case of food-borne illnesses in which milk or a dairy product is implicated or closely associated will change public sentiment and lead to further legislative control and limitations. Some states have taken interesting approaches to try to balance the food rights of its population with the concerns of health officials—either through shrinking the "footprint" of the sales of raw milk through allowing only on-farm sales or limiting production by volume or herd size or through requiring or providing milk quality testing.

In chapter 5 you learned about some of the animal illnesses that can be transmitted through unpasteurized milk and that were in fact a leading cause of illness in the early parts of the twentieth century. Some of these diseases, particularly tuberculosis and brucellosis, have been eliminated in many states. These states are even labeled "brucellosis- or tuberculosis-free." But that doesn't actually mean that the diseases don't exist anymore in those states. Wild populations of elk and bison are of particular concern for transmission to domestic animals. (By the same token, domestic livestock can pose risks to unexposed wild animals.) Therefore, even a disease-free state may require animal testing, especially when the milk is meant for unpasteurized consumption by humans. There are new diseases of concern on the horizon and should be of concern to all farmers. Unfortunately, regulators have a history of aggressive, preemptive action that may or may not prove necessary. I encourage every farmer to be aware of current health issues and concerns, and do your best to reduce the risk of transmission through herd health and biosecurity (for more on biosecurity see sidebar in chapter 6).

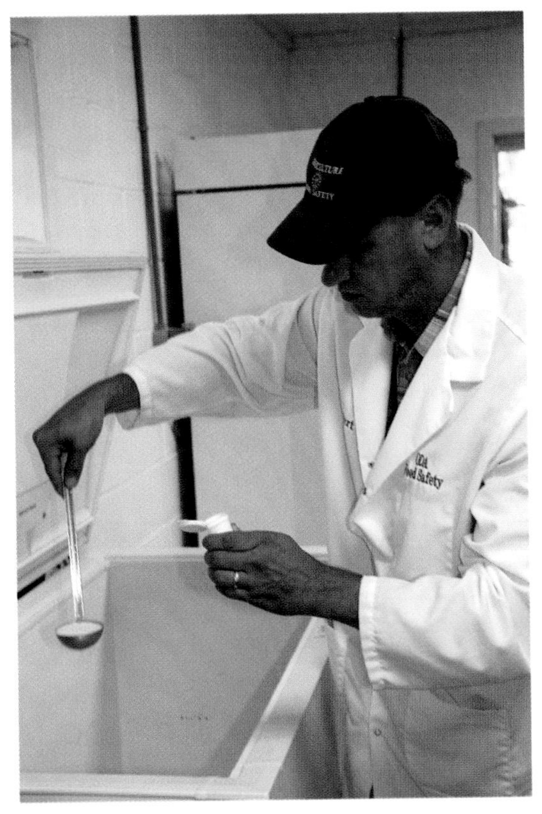

Licensed producers typically have their milk sampled and tested by a regulatory agency. Here an Oregon Department of Agriculture inspector takes a sample of milk for testing.

CAN PASTEURIZED MILK REALLY CONTAIN THE SAME BACTERIA COUNTS AS RAW MILK?

Yes . . . and no! An often-cited statement is that pasteurized milk can legally contain the same total bacteria counts that are also achievable by raw-milk producers. This statement can be both true and misleading. First, it is true that high quality raw milk bacteria counts can easily be lower than the limit allowed for pasteurized milk. However, the numbers allowed in pasteurized milk are not a count of the same type of bacteria as found in raw milk. Remember that pasteurization does not sterilize milk; certain heat-tolerant, or thermoduric, bacteria and bacteria that form protective spores can survive. These are typically spoilage bacteria, not illness-causing pathogens. These regulatory acceptable bacteria levels in pasteurized milk are meant to reflect the presence of this type of bacteria. Remember that dairies collecting milk destined for pasteurization rarely if ever follow the same strict process steps that the raw milk producer should. The rather high bacteria counts allowed in milk intended for pasteurization reflect this.

WHAT THE HECK IS A CFU?

To understand your lab results and speak with confidence about the bacteria counts in your milk, you should understand the difference between a cell, a colony-forming unit (CFU), and a colony. Remember that most bacteria reproduce by cell division, not by sexual reproduction.

There are basically two different ways to count bacteria in a sample: first, through direct microscopic observation in which all cells, living and dead, are counted and, second, by placing a sample on a growth medium, giving it the optimal conditions for growth, then counting the clusters of cells (colonies) that result. Colonies can be produced by any number of original, living cells, from one to several. The term "colony-forming unit" refers to whatever cell presence produced a colony. It is not a true count of the actual number of cells; it shows the viability of those cells in the sample.

Licensed facilities, whether they be Grade A or in another category, likely will be tested by their regulatory agencies for milk quality. Standard tests usually include bacteria counts, somatic cell counts, and antibiotic residue testing. Where funding is available, additional tests may be included. In some states with regulations allowing the sale of milk from small, unlicensed producers or through herdshares, there is also often testing done or required. Typically, tests will include a standard plate or aerobic count, which counts all bacteria, both "good" and "bad." Somatic cell count also is done—with federal legal limits being different for cows and goats (sheep being the same as cows). Some states, however, may set lower limits. An antibiotic residue test is always standard, and many states also include a total coliform count. Beyond these basic tests, though, there are many others that I highly recommend that raw milk producers periodically have done on their milk. Remember that many of the same tests performed by the regulatory agency also can be done on the farm, and although not "official" they can greatly assist with the monitoring of the farm's production standards. We'll cover many of these do-it-yourself tests later. Finally, remember that it is your family and friends, your farm, your reputation, and your livelihood on the line—not that of the regulators!

Talking the Talk: Understanding the Lingo of Lab Milk Tests

When we first became a licensed dairy, I thought that when the state took milk samples (in Oregon they take them every six weeks), did the lab work, and sent the results that our inspector would tell us if the results were less than ideal. But no. In fact, the inspector saw the results only if the sample exceeded the allowed level. Once I realized that, it became important to understand what the entire spectrum of results meant—what was ideal, what was not—as well as how to improve the results. Fortunately for me, the Dairy Practices Council (see the bibliography) has a couple of handy little booklets on the topic. Table 10.1 is a quick overview of common lab tests for milk. We'll go over each test in more detail and explain what the results mean, including possible causes.

TABLE 10.1 Milk Test Result Ranges*				
Test	**Ideal**	**Acceptable**	**Common Industry Goals****	**Federal Regulatory Limit for Grade A Milk (prior to pasteurization)**
Raw SPC or APC, standard or aerobic plate count: counts total numbers of bacteria in sample	< 1,500 CFU/ml	< 5,000 CFU/ml	< 10,000 cfu/ml	100,000 CFU/single producer 300,000 CFU/commingled tanker
PIC, preliminary incubation count: counts the number of psychrotrophic (cold loving) bacteria in a sample	< 3–4 × the raw SPC/APC	< 25,000 CFU/ml	25,000–50,000 CFU/ml	No legal limit
LPC, lab pasteurization count: counts the number of thermoduric (they survive pasteurization) bacteria in a sample	<250–300 CFU/ml	Less than ⅓ of raw SPC		(See the sidebar "Can Pasteurized Milk Really Contain the Same Bacteria Counts as Raw Milk?" for more on pasteurized allowable counts) 20,000 CFU/ml
Coliform count: counts the total number of coliform bacteria in a sample	< 10 CFU/ml	< 50–100 CFU/ml		< 10 CFU/ml (pasteurized milk)
SCC, somatic cell count: cows, sheep	Varies by herd		About 200,000	< 750,000
SCC: goats	Varies by herd		About 750,000	< 1,500,000
***E. coli* and 0157:H7 specific counts:** tests that tell the difference between general coliforms and *E. coli* or pathogenic strains	Zero (often read as < 1 CFU/ml)			Zero (often read as < 1 CFU/ml)
Listeria and L. monocytogenese	Not found (often read as < 1 CFU/ml)			Not found (often read as < 1 CFU/ml)

** Source: Dairy Practices Council*
*** Related to bonus payments made to the producer by the milk processor*

Raw Standard Plate Count (SPC) or Aerobic Plate Count (APC): This basic test, usually incubated at 90°F (32°C) for 48 hours, gives a count of the total number of colony-forming units of all bacteria in the sample that will grow at that temperature. Because these counts are so high compared to other, more specific tests, the milk sample usually is diluted to 1 part milk to 100 parts water (1:100). If the count is high, a dilution of 1:1000 can be done. Some labs do the higher dilution as their standard test. The results, however, always are given in cfu per milliliter of milk. Although this test counts both "good" and "bad" bacteria, higher-than-ideal counts could indicate the presence of a higher-than-desirable number of pathogens and spoilage bacteria.

Preliminary Incubation Count (PIC): This test rarely is required or performed by regulators but can be a regular part of milk quality testing done by a certified lab. The PIC tests for the number of psychrotrophic (cold loving) bacteria in a sample. The milk sample is held at 55°F (13°C) for 18 hours; then an SPC is performed. High PIC is often indicative of poor cooling in addition to the causes listed in table 10.2. If milk is chilled too slowly, it provides a good environment for bacteria that like to grow at just above refrigeration temperatures.

Laboratory Pasteurization Count (LPC): Also known as thermoduric bacteria count, this test is performed after the milk sample is heated to 145°F (63°C) for thirty

TABLE 10.2 Likely Causes of High Bacteria Counts

Test Result	Natural Bacteria	Mastitis Bacteria	Dirty Udders	Dirty Equipment or Contaminated Water	Poor Cooling
SPC > 10,000	Doubtful	Possible	Possible	Possible	Possible
SPC > 100,000	Doubtful	Possible	Doubtful	Likely	Likely
LPC > 300	Doubtful	Doubtful	Possible	Likely	Doubtful
PIC high, 3–4 × SPC	Doubtful	Doubtful	Possible	Likely	Likely
SPC high > PI	Doubtful	Presumed	Doubtful	Doubtful	Doubtful
Coli Count > 100	Doubtful	Possible	Likely	Possible	Doubtful

Source: "Bacteria Counts in Raw Milk," Richard L. Wallace, University of Illinois Extension publication; http://www.livestocktrail.illinois.edu.

minutes to simulate batch pasteurization; then an SPC is performed. Thermoduric bacteria, which normally survive pasteurization, can then be counted. Very few, if any, of these bacteria are disease causing, but they do cause flavor and spoilage problems. The federal legal limit of 20,000 CFU/ml of bacteria in pasteurized milk is meant to accommodate the possible presence of these bacteria.

Coliform Count or Total Coliforms: Although most coliforms are harmless, they are also an indicator organism, meaning that their presence can indicate that a certain number of pathogenic coliforms also might exist in the sample. (Think of it as "guilty by association.") The milk sample is plated on growth medium and incubated at 90°F (32°C) for twenty-four hours. Because coliforms grow faster than most bacteria, their presence is seen sooner, so the test is quicker. If total coliform count is high, a separate *E. coli*-specific test can help differentiate between environmental versus fecal coliforms. However, if *E. coli* 0157:H7 is to be detected, a special plate is required, since this strain of coliform is not detected by the same means as other *E. coli.*

Somatic Cell Count (SCC): The somatic cell count of milk is indicative of the health of the animal's udder. As you'll recall, leucocytes (white blood cells) and other body tissue cells in milk are referred to as somatic cells. Although some books and websites refer to somatic cells in milk as "pus," this is an inaccurate and misleading description. Pus, the weeping or oozing fluid from an infected wound, does contain white blood cells, but it is directly associated with the body's response to an infection or foreign body (such as a sliver). Milk in the healthy udder will contain some white blood cells, but they are not necessarily the result of an infection. In chapter 4 you learned of other differences in the way that cows, goats, and sheep secrete milk, which also lead to hugely different SCCs. In addition, many other factors will influence SCC, including the age of the animal, the number of times she has freshened (given birth), and even the season (with shorter daylight hours correlating with increasing SCC).

The Mad Scientist—How to Be an Amateur Lab Tech

If you do your own lab tests, the results will not be recognized officially by any regulatory agencies, but they still will be a powerful tool in your efforts to produce high-quality milk. In fact, the low-cost, quick results, and the lack of involvement of government agencies (currently positive listeria tests done by certified labs must be reported to the FDA) makes regular, on-farm testing appealing. I have used our testing program to verify the efficiency of workers (ourselves and our kids), test out new cleaning solutions to see if they are doing the job, and pinpoint problems in equipment. We'll start with simple observations of milk, then go on to how to check for elevated somatic cell counts, how to test for antibiotic residue, how to do a simple lactic fermentation test to observe for gas-forming bacteria, and finally, how to perform the ultimate in bacteria testing (usually the same method used by certified labs), 3M Petrifilm Plates.

Milk Abnormalities You Can Detect with Your Senses

You can detect many flaws and problems in milk by simply using your senses. Most of the flaws that you can see, smell, or taste are not indicative of the presence of pathogens but are still important to find and correct for the production of high-quality fluid milk.

Visual Flaws

In chapter 5 you learned that milk should be observed at a minimum of two points during the milk collection process: first when the first few squirts, the foremilk, are milked from the animal into a strip cup, which often includes a strainer, and second, after the milk is filtered (you should inspect the filter). During both of these observation times, you should look for milk abnormalities such as clumps, flakes, crystals, strings, and blood. All of these are indicative of a problem in the animal's udder. If you observe any of these, you should perform a test for somatic cells immediately, such as the California Mastitis Test (CMT). If these abnormalities are caused by an udder infection, it is possible that bacteria that can cause a human illness are present, so observation is one of the first and most proactive steps you can take in the production of high-quality milk.

Flavor and Aroma Flaws

Desirable taste and smell are subjective goals for the milk producer. For example, I personally do not care one bit for the flavor of milk from animals fed on mostly pasture grasses; milk from animals that are fed a mixed diet that includes dry fodder and a bit of grain is much more to my liking. Individual animals also can have consistently more or less pleasant tasting milk, either from their own metabolism, food preferences, and even homeostatic imbalances. Nonetheless, some flavors and smells are considered flaws. Here are some general guidelines for troubleshooting these off-flavors and aromas:

Spoiled or unclean. Characteristics: dirty-animal or barnyard taste, spoiled fruity flavor or aroma. Possible causes: contamination or bacterial growth during storage. Likely source: dirty equipment or udders, psychrotrophic bacteria.

SMALL DAIRY PROFILE: FIORE DI CAPRA, ARIZONA-LICENSED GOAT'S MILK DAIRY

Alethea Swift, herdmistress and head dairymaid at Fiore di Capra in Arizona, makes delicious goat's milk yogurt, cheese, ice cream, raw kefir, and bottled raw milk. I was lucky enough to sample some of the farm's incredible yogurt as part of my "research" for an article I was writing for *culture* magazine. Licensed to produce cheese and milk for commercial sale since 2006, Fiore di Capra takes advantage of product diversity to provide a sustainable income to the small family farm. Her herd of around a hundred La Mancha goats produces milk year-round, defying the usual seasonality that affects the income stream of most goat's milk producers.

Alethea and her husband Michael began their business intending to make only cheese but quickly saw the advantages in bottling and selling a portion of the milk to provide quick income and answer a demand that she saw in the market. The farm currently sells about 100 gallons of raw goat's milk per week, packaged in half-gallon and 1-quart containers for $8 and $5, respectively. Her product also appears in retail stores, as Arizona law allows. Swift says that she has been gratified not only by the income, but by the appreciation customers continually express at the quality of her product and the benefits it has brought to their lives.

Arizona tests the milk monthly, and Swift also performs routine Petrifilm plates for bacteria and coliforms right at the farm. In addition to state health requirements, Fiore di Capra also tests for Q fever, Johnes, and mycoplasma. While of these only Q-fever is a known human health hazard, Alethea is conscientious of the importance that overall animal health plays in the production of high-quality milk. Arizona requires that milk be tested for antibiotic residue, even when the production is 100 percent farmstead (meaning that milk is produced only by the animals owned by the cheesemaker or milk producer). To accomplish this Swift utilizes the Delvo P Mini Test system.

Alethea shared with me that her inspectors have been wonderful to work with—enforcing regulations while at the same time supporting the producers. The farm's location in Arizona gives it access to year-round farmers' markets, helping balance the books with a steady source of sales. Swift's recommendation to those considering becoming raw-milk producers is to learn as much as they can about the legalities of the market and the factors affecting quality. If Fiore di Capra's rich, creamy Greek-style yogurt is any indication, Alethea Swift has succeeded in taking her own advice.

Learn more about where to find Fiore di Capra products at *www.goatmilkandcheese.com.*

Alethea Swift uses a refurbished milk bottler at her small dairy, Fiore di Capra, in Arizona.
Photo courtesy Michael and Alethea Swift

Rancid, bitter. Characteristics: soapy, baby vomit, blue cheese flavor or aroma. Cause: breakdown of milk fat. Likely reasons: excess agitation or foaming of milk during milking, late lactation (high SCC), fluctuation of temperature during cold storage, all leading to damage of the fat globule.

Malty, acidic. Characteristics: Grapenut cereal-like or sour flavor or aroma. Cause: bacterial growth during storage. Likely reason: inadequate or too-slow cooling of milk.

Oxidized. Characteristics: cardboardy, old-oil flavor or aroma. Cause: oxidation of butterfat. Possible reasons: excessively high-fat feeds, low levels of vitamin E, presence of certain metals (copper contamination, rust), and exposure to light.

Feedy. Characteristics: herbal, unnaturally sweet or spicy flavor or aroma. Likely cause: odors and flavors consumed by animal and transmitted to milk.

TIPS FOR ANALYZING MILK FOR FLAVOR AND AROMA FLAWS

1. For maximum aroma analysis, pour sample into a jar, leave a headspace, close tightly, and warm to about 60°F (16°C) (in a water bath). Open the jar directly under your nose, and inhale deeply.
2. Compare fresh milk to milk that has been stored for several days to compare differences in aroma and flavor.
3. If necessary, compare samples from each animal to rule out individual variations in flavor and aroma.

Testing for Somatic Cells

To review quickly, somatic cells always are present in milk, to one degree or another, depending on the species, the breed, and factors such as the stage of lactation (how long since the babies were born) and the time of year. Somatic cells are not bacteria; they are white blood cells, leukocytes, and other parts and bits of the cells that line the inside of the udder. The legal limits for high-quality milk are far higher than those of a normal, healthy animal and should not, in my opinion, be used as the limits for any herd if you are focused on udder and animal health.

Just what is an ideal somatic cell count (SCC)? The short answer is, "It depends on your herd." The more complete answer involves frequent monitoring of SCC by the herd manager, coupled with observation of milk and milk production. Over time and with the collection of data, the herd manager can decide what is normal and acceptable. When SCCs are higher than they should be, the udder tissue will suffer damage from both the infection and the increased movement of blood components through the tissue into the milk—and milk production will drop. This kind of change quite often will occur without any visible changes in the milk! So don't count on the strip cup to tell you that there is a problem.

There are several kits and tests available for checking SCC in the milking parlor (but don't forget that you can do these tests at any time on any milk). The gold standard remains the California Mastitis Test (CMT), which uses a divided paddle into which an equal amount of the purplish reagent is combined with milk from each quarter of the udder (or half if from goats or sheep), then swirled for a few seconds. If any thickening of the liquid occurs, it indicates a higher-than-desirable somatic cell count. Although the test is subjective, meaning the reader has to do a bit of guessing, it is quick, inexpensive, and fairly accurate. Here is a guide for interpreting the results:

The California Mastitis Test, or CMT, will show visible viscosity changes when high numbers of somatic cells are present; these results are acceptable.

Trace gelling = 300,000 cells/ml
Thickening but not clumping = 500,000–1,000,000 cells/ml
Thickens and clumps = > 1,000,000 cells/ml
Clumps and sticks to paddle = > 2,000,000 cell/ml

I recommend that a CMT be performed on every animal in the herd each month, unless another method of SCC is regularly being done (such as DHI—see chapter 8 for more). Even if SCCs are being monitored, through either individual or bulk milk tank samples, it is a good idea to have a CMT kit on hand to check an individual when her SCC is slightly elevated. This way, you can determine if only one part of the udder is affected, then take holistic measures to help the udder heal, long before antibiotics might be needed. Here at our farm we have never—knock on wood—had an acute case of mastitis, thanks to vigilant SCC monitoring.

Antibiotic Residue

Most small dairy farmers do not use antibiotics and antibacterials routinely. When they are needed it is relatively simple to identify treated animals with a marker or other identification means and make sure that no milk from treated animals is used until their milk is free from any traces of the drugs. This is called "withholding time." It is critical that you develop policies on how to identify animals you are treating with antibiotics and how you will manage them to be sure their milk stays segregated from the milk intended for human consumption. In addition, milk with antibiotic residue should not be fed to animals, including calves, kids, and lambs, as it exposes them to subtherapeutic levels (meaning amounts that are not strong enough to combat the condition or disease that they were originally designed to fight) that will contribute to antibiotic resistance

The more people that are involved with management of the dairy herd and milk production, the more likely it is that a mistake will occur and milk with antibiotic residues will make it through to the human supply. To try to prevent this, regulators may recommend or even require that larger dairies (and even some smaller facilities) perform an antibiotic residue test on each batch of milk they sell. There are several types of antibiotic residue tests that can be done on the farm. They vary in complexity, completion time, and cost—both of the initial setup and for each test. If you are required to test for residues or feel that it is important, I recommend contacting a supplier of such tests, such as Nelson-Jameson (see appendix A), and finding out which test is best for your farm.

Lacto-Fermentation Tests

Although most of us are familiar with the term "lacto-fermentation" as it relates to the fermentation of foods, including turning vegetables, milk, and meats into tasty new products such as kraut, cheese, and salumi, lacto-fermentation also can be used as an inexpensive—actually, almost free—milk-quality test. This form of testing is far more common in Europe, but I hope to foster its increased use by both fluid milk producers and cheesemakers. In a nutshell lacto-fermentation of milk to test its quality involves holding milk samples at about 95°F (35°C) for twenty-four to forty-eight hours, then observing the sample for changes. Although this test is a bit more use to the raw-milk cheesemaker, it still can be quite useful and incredibly revealing about milk quality to the fluid milk producer. While I was in England, I saw this done using test tubes set into a tiny laboratory water bath, but I have not been able to find such a piece of equipment here in the United States. Instead, you can use the same incubator that I recommend from Nelson-Jameson or you can build your own from a small ice chest and heat source such as a thermostat controlled heating pad for reptile habitats.

I found an incredible article online called "Assessment of Milk Quality and Dairy Herd Health under Organic Management" (see bibliography) about an in-depth study in which lacto-fermentation proved a reliable method for monitoring milk quality. Even if you are doing a more quantitative test such as Petrifilm (we'll cover that next), the lacto-fermentation method provides a tangible, "real" result that many people will be able to relate to better.

Lacto-fermentation can reveal poor milk quality without fancy lab tests. This milk reveals "D" quality, not fit for making into cheese or drinking raw.

What You'll Need

- Baby food jars or test tubes
- Incubator or water bath

Steps

1. Sterilize the jars and lids.
2. Fill each jar with milk two-thirds of the way full. Screw on the lid, making sure it is not completely tight.
3. Incubate the samples for twenty-four to thirty-six hours at 90 to 95°F (32–35°C).
4. Observe the samples for visual indicators (see below).
5. Note the aroma.
6. Note the taste. (If the sample is less than B quality, tasting is not recommended.)

According to the article I mentioned earlier, the results of lacto-fermentation can be divided into four results: Types A, B, C, and D milk, with A being the best. The results are defined in the article as follows:

- A: Homogeneous curd, 40 to 50 percent, plus whey.
  Best quality, suitable for fine aged cheeses and raw fluid milk.

- B: Homogeneous yogurt-like gel, 100 percent.
  Contains low levels of enzymatic reaction inhibitors. Inherent to some breeds and individuals, or from sanitizer or antibiotic residue.
  Okay to use in processed products.
- C: Liquid milk.
  May indicate excessive sanitation.
  Incubate longer and see if it turns to A, B, or D milk.
- D: Limited curd, 10 percent of the volume, or flaky, broken-down curd.
  Feed problems, mold, yeasts.
  Not acceptable for use.
- Other possibilities: Milk with "too much" lactic bacteria might ferment too rapidly and become D milk.
- Curd with gas = coliform contamination.

Peter Dixon, owner of Dairy Foods Consulting in Vermont, also recommends using lacto-fermentation to determine the presence of gas-forming bacteria such as coliforms. Even if you don't plan on using this form of testing regularly, I encourage you to pop a couple of samples in the incubator and see what happens. It is truly enlightening!

Petrifilm: Grow Your Own

In the "old days" lab technicians would grow bacterial cultures in glass containers called petri dishes or plates. (Many still do). Then along came the company 3M with a disposable product called Petrifilm. These plates take the guesswork (and the skill) pretty much out of the picture and allow the technician simply to inoculate the plate with the sample, pop it in an incubator, set a timer, then count the results. Nelson-Jameson sells a very inexpensive incubator (under $100).

So with a few supplies and a good set of directions (and a pair of glasses and good lighting if you are over fifty like I am), all of us can be our own amateur lab tech. Remember, you are growing bacteria, likely no more than you find any place on your farm, but you still should take precautions against contamination. You should perform the tests away from your milk processing area, and you should wear gloves when you

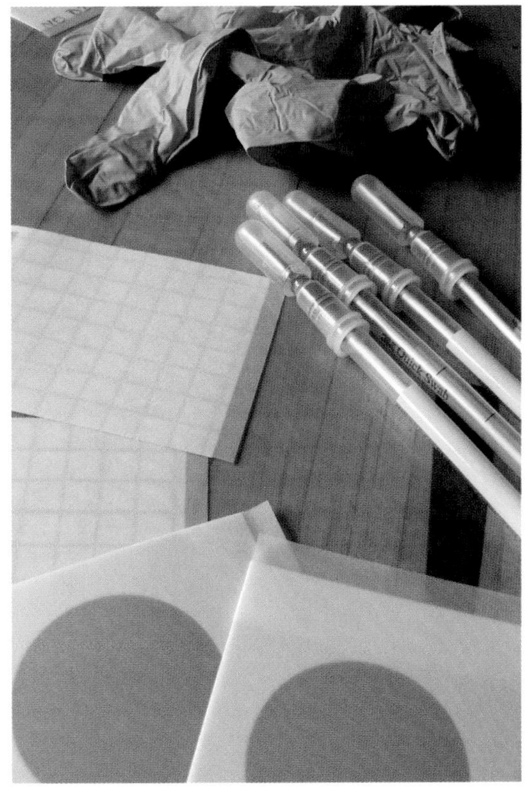

A small incubator, Petrifilm plates, and a few other supplies are all that you need to get started testing milk for bacteria.

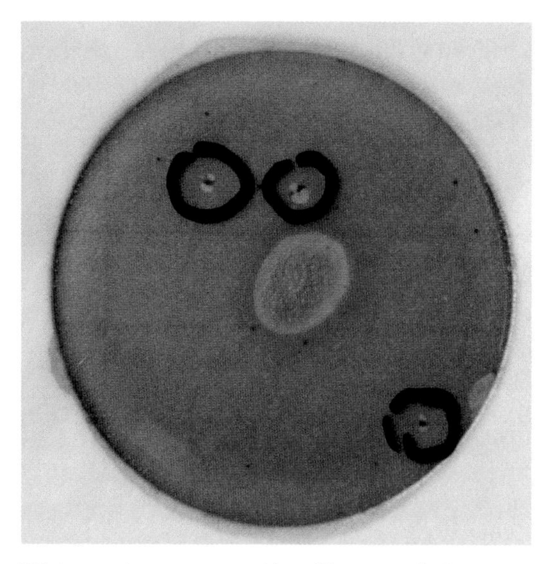

This image shows an acceptable coliform count (colonies circled in black ink) showing 3 cfu/ml of milk.

are reading the incubated plates. Plates are destroyed in the lab using an autoclave. It has been recommended to me by two industry quality-control specialists that soaking the finished plates in 190 proof alcohol or straight bleach, or heating in a small autoclave (or pressure cooker) are reliable ways to sterilize the plates before disposal. I use the 190 proof alcohol method.

There are many types of Petrifilm plates that you can purchase, but unless you are going to set up a true laboratory with biosecurity and proper disposal capability, I recommend sticking with a few simple tests that don't encourage the growth of some nasty bugs. Aerobic plate count (APC) and coliform count plates should be first on your list. Quick Swabs for testing surfaces such as milk pails and bottling equipment are also a good idea to have and use. Store your Petrifilm plates in a cool, dry area or as directed on the packaging, and be sure to seal the individual film packets tightly. They are so sensitive that they simply can be exposed to the air and culture contaminants (with the cover film peeled back) via that route. You don't want to expose them until ready to inoculate.

What You'll Need:

- Incubator
- Petrifilm plates—aerobic count and coliform count plates (optional *E. coli* plates)
- Plate spreader (comes in each box of plates)
- 1 ml syringe or pipette
- Fine-point black marker
- Milk sample

Step-by-Step Instructions for Plating a Milk Sample for APC and Coliform Count

1. Obtain a 1 ml sample of milk using a sterilized 1 ml syringe or a pipette. (You may need to dilute the milk to read the results more easily when performing an APC.)
2. Lift the film on the room-temperature Petrifilm plate, and place the sample in the center. Lower the film gently.
3. Center the plate spreader (smooth side up for APC and dimpled side up for coliform count) over the sample. Lower it onto the film, and press firmly to spread the sample in an even circle.
4. Place the Petrifilm in the incubator at 90°F (32°C). (Note: The compact incubator from Nelson-Jameson comes with instructions

that state that the shelf temperature is 10 degrees F lower than the thermometer readout, so adjust your temperature accordingly and consider periodically verifying the temperature with a calibrated thermometer.)

5. Incubate:

   APC: 48 hours
   Coliforms: 24 hours
   E. coli: 24–48 hours

6. Remove the plate from your incubator. Using a fine-point Sharpie pen, count each dot using the pen to mark as you count (so you don't double-count any cfus).

   APC: Count each red dot.
   Coliform count: Count each red dot in or directly next to an air bubble.
   E. coli count: Count only the dark blue/purple dots associated with an air bubble.

7. If the plate has very few dots, count the entire plate. If there are quite a few, you can count one square and multiply the result by 20. Do this with several squares so you get an accurate average. (Each square represents 1 square centimeter, and the plate area is 20 square centimeters; thus the multiplication by 20.)
8. Dispose of the plate properly. In certified labs this is done through sterilizing in an autoclave or incineration. In the home lab it may be adequate to soak the plate in 190 proof alcohol or straight bleach, sterilize it in a small autoclave or pressure cooker, or incinerate it. If you are incubating only for total bacteria and coliforms, not selectively encouraging the growth of specific pathogens, these plates are of relatively low concern.

If an undiluted sample grows too many cfus, it is impossible to get a good count, since the plate will be crowded with overlapping colonies. Not only do the colonies

WHEN TO USE *E.COLI*-SPECIFIC PETRIFILM

When total coliform counts exceed the desired levels—desired level is less than 10 cfu/ml—it is a good idea to run an *E. coli*–specific plate. This will help you determine if any of the coliforms are from fecal sources. It does not differentiate between different strains of *E. coli*, such as 0157:H7, but the goal still should be "none present." When we had some high coliform counts one fall, over 400 cfu/ml, the *E. coli*–specific plate sample showed zero *E. coli*. Although this was somewhat reassuring, we still had some work to do to troubleshoot and solve our high counts! *E. coli*–specific plates can be used to count total coliforms, which will appear as they do on the regular coliform count plate, as a red dot in or right next to an air bubble, while *E. coli* will look the same but will be dark blue/purple in color. You might choose to run these instead of the total coliform plates, even though they are a tad bit more money; peace of mind is valuable!

form too closely, but they can compete with each other for the growth medium on the plate, possibly leading to falsely limited growth. If gas-forming coliforms are present, the air pockets they create can lift the film away from the plate and give erroneous readings. You carefully can dilute the sample by 50 percent with sterile water, then multiply the resulting count by 2. On our farm we do not dilute the sample but conclude that if it is too crowded, the milk is not good enough to use—a qualitative (quality based) versus a quantitative (quantity based) result.

Swabbing

You can find gaps in your food safety program quickly by testing food-contact surfaces and other surfaces that could contribute to the cross contamination of your product. This spring we made some changes in our general procedures that led to increased bacteria counts in our milk. By using swab testing we were able to pinpoint where the problem was originating before the bacteria levels got too high.

Swab Testing: What You'll Need

- Incubator
- 3M Petrifilm plates: aerobic plate count and coliform count
- Plate spreader (comes with Petrifilm plates)
- 3M Quick Swabs
- Fine-point black marker

Once you have gathered your supplies, you can begin taking samples. It is a good idea to test far more surfaces in the beginning of a testing program than you may need to do on follow-up tests—this will help establish a baseline of awareness. Once enough tests confirm that cleaning protocols are effective, you may be able to decrease the number of surfaces you test, as well as the frequency. But any time you establish a new protocol, it is a good idea to increase the swab testing to verify that the new protocol is working.

Step-by-Step Instructions for Successful Swab Testing

1. Using a Sharpie or other marker, write the source of the sample and date taken on the Quick Swab container.
2. When you are ready to swab the surface, bend the neck of the liquid-filled end of the Quick Swab so the nutrient broth contained in the bulb flows into the end that contains the swab. Squeeze the bulb so all the solution is drained.
3. Twist the top of the tube, and remove the swab from the tube. Hold the tube so the broth solution remains inside once the swab is removed.

4. Rub the end of the swab, holding it at a slight angle so the sides make some contact with the surface, on the desired area to be sampled. Rub the swab three times over an area of roughly 3 to 4 square inches.
5. Return the swab to the broth, and close the tube.
6. Shake the tube for about ten seconds to mix the sample into the broth.
7. Remove the swab from the tube, squeezing it inside the neck of the tube to remove as much of the solution from the absorbent material as possible.
8. Peel back the film on the sample plate (APC or other), and carefully pour the solution onto the center of the plate. It tends to run out very quickly, so this step is tricky to do properly.

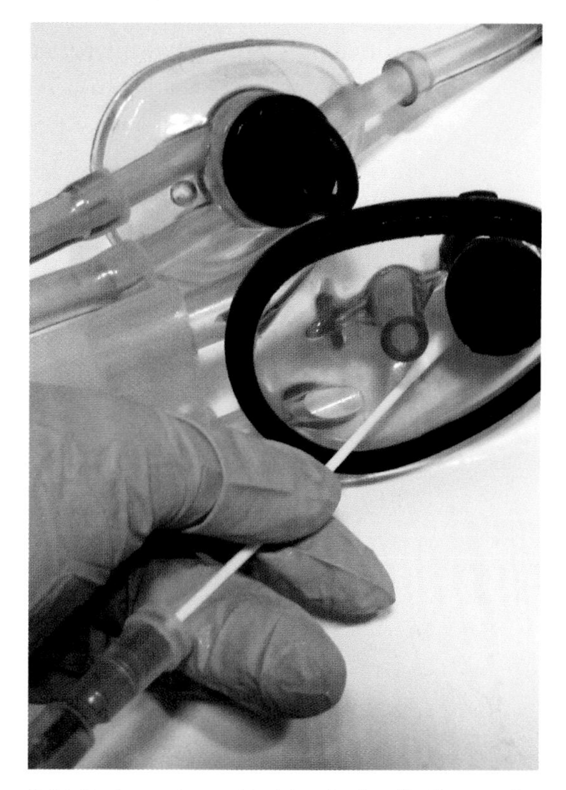

Quick Swabs can be used to determine the effectiveness of an equipment-cleaning solution quickly.

9. Use the plate spreader to press the sample gently into the plate. Use the flat side for coliform plates and the recessed side for APC plates.
10. Allow the plate to sit for about one minute so the liquid sample will gel with the plate.
11. Incubate as directed for the type of sample being run (for APC it is 90°F [32°C] for forty-eight hours; for coliform plates incubate at 90°F for twenty-four hours).
12. Follow steps 6-8 for Petrifilm plate test on page 146.

Handling Less than Great Test Results

So what do you do when the state lab or your own testing reveals less than gold star results? Table 10.2 suggests the most likely causes of different bacteria count problems in raw milk. But as often as not, more than one source is contributing to the issue. Troubleshooting high bacteria counts is a bit like a game of Sleuth—and you often have to be a persistent detective to find the cause quickly.

To put it simply, any breakdown in the six goals for quality-milk production can lead to troublesome levels of bacteria in milk.

To put it simply, any breakdown in these six goals can lead to troublesome levels of bacteria in milk. So let's review them:

Six Goals for the Control of Quality Milk

1. **Healthy animals** = homeostasis that fosters proper milk enzyme balance, proper milk nitrogen (milk urea nitrogen—MUN) content, udder health (lower SCC), resistance to disease, and more.
2. **Clean animals** = clean milk collection and clean air entering the milking system.
3. **Clean teats** = limited unwanted bacteria and other contaminants entering milk during collection.
4. **Clean parlor** = clean air quality entering milking system.
5. **Clean equipment** = surfaces free from any contaminants—bacterial or chemical—that could enter milk supply. This includes ensuring that water used to rinse equipment is free from contaminants.
6. **Proper chilling** = rapid chilling to and maintenance of refrigeration temperatures to prevent the growth of microorganisms.

When you begin troubleshooting, start by evaluating the effectiveness of each of these six goals. If the microbiological change in your milk is sudden, it is probably not related to goal 1, healthy animals. In fact, this goal should be evaluated in an ongoing, constant fashion, since the factors that disrupt homeostasis usually occur over time. The other goals, or controls, can go by the wayside in an instant.

When you troubleshoot bacteria-count problems, you simply are questioning the effectiveness of all the things you do that are meant to prevent the problem.

To best troubleshoot problems, you first have to know how to achieve the goals, or controls! Not only do you have to know, but everyone performing the tasks related to the goals also must know the right steps to attain them. These steps are what are known in the industry as Standard Sanitary Operating Procedures, or SSOPs. In the next chapter we'll go into the development of a proper risk reduction plan, but for right now I want you to start thinking about the goals and what you do to achieve them. When you troubleshoot, you simply are questioning the effectiveness of all the things you do to achieve the goal—the goal that according to the bacterial problem is most likely not being met.

· 11 ·

Risk Reduction Plans

In this chapter you'll learn how to apply the science and practices you learned in the preceding chapters to a written plan for raw-milk production. Though this is geared more for the commercial producer, the home dairy will benefit from the same process knowledge that these plans add to the production of high-quality milk. These policies are still often referred to as "food safety" plans, but they more often are correctly described as "risk reduction" plans. This title better acknowledges that, no matter how good the plan is, no food can be guaranteed 100 percent safe. When considering raw-milk production, to me it boils down to creating the best possible risk reduction program, while still acknowledging that raw milk is a worthy product, even if it might be considered higher risk than its heat-treated cousin. (Other foods people consume regularly, such as raw leafy greens and raw oysters, fall into this category as well.) I will use the term "food safety" here and there through this chapter, but remember, while 100 percent safety might be the goal, no food can ever be guaranteed 100% safe for everyone. Your responsibility is to reduce the risks to the lowest level possible.

The most common approach to risk reduction in the United States and Europe is called Hazard Analysis and Critical Control Points or, more commonly, HACCP. Although this program was developed by NASA, creating and following a HACCP plan isn't rocket science! I'll cover the key principles of HACCP and walk you through the steps of creating your own risk reduction plan. You don't have to, and maybe even shouldn't, call a plan you create on your own, a HACCP plan. I encourage you to think of it as a process documentation plan—and if you *really* know the right steps in the process, you will reduce the risks associated with production.

If you are a consumer of raw milk produced by others, consider it your right to understand these principles and perhaps even demand them from your producer. If you are a producer, it is your reputation, conscience, and livelihood that are on the line, and if you drink the milk you produce it's your health!

Instead of calling it a HACCP plan, think of it as a process documentation plan—and if you *really* know the right steps in the process, you will reduce the risks associated with production.

What Is This Thing Called HACCP?

HACCP (pronounced with a short "a" as in the word "at," with a soft "c": "ha-sip") is a system designed to define every step of a product's manufacture, identify the points in the process that are critical for food safety, and create a plan that delineates and documents safe manufacturing processes. It is a preventive program, focusing on the process rather than on product testing, although testing can be an important part of a good plan. In the United States HACCP programs are mandatory in certain food industries, and new regulations (brought about through the Food Safety Modernization Act), due to take effect soon, may require far more producers to maintain a risk reduction plan of some sort. Larger dairy processors often are required by their retail clients to have HACCP plans in place that are routinely monitored and certified by an outside third party, a person or organization trained for such work.

A HACCP plan starts with a foundation called prerequisite programs. This term covers good manufacturing practices (GMPs), standard sanitation operating procedures (SSOPs), and standard operating procedures (SOPs). Don't let these long terms and their acronyms throw you for a loop; they can be summed up as sensible policies that help make good food and the step-by-step instructions of how to do it. Other parts of the prerequisite program include the development of policies that cover issues that must be addressed for the GMPs to be fulfilled. Eight of these policies are mandated by the National Conference of Interstate Milk Shippers, but several more also are recommended for producers.

Understanding GMPs is the foundation for creating good policies and the steps needed to ensure that those policies are followed.

In the next part of this chapter, I'll walk you through all of these prerequisite or foundation programs and how to create them for your facility. Once the foundation programs are in place, you can perform the HACCP portion of the plan. Because the actual HACCP portion comes last, I find that calling the entire program a HACCP plan is a bit misleading. I like the term "risk reduction plan" much better and think it is more accurate.

THE SEVEN PRINCIPLES OF HACCP

There are seven official principles of HACCP as defined by the National Advisory Committee on Microbiological Criteria for Foods (NACMCF).

- Conduct a hazard analysis.
- Determine critical control points.
- Establish critical limits.
- Establish monitoring procedures.
- Establish corrective actions.
- Establish verification procedures.
- Establish record-keeping and documentation procedures.

Building a Complete Risk Reduction Plan in Seven Simplified Steps

I have broken the building of this into seven steps that made sense to me. Always keep in

MY RECOMMENDED DAIRY PRACTICES COUNCIL GUIDELINES FOR THE SMALL-SCALE MILK PRODUCER—AVAILABLE FOR PURCHASE AT *WWW.DAIRYPC.ORG*

- DPC 8, Good Manufacturing Practices for Dairy Processing Plants
- DPC 9, Fundamentals of Cleaning & Sanitizing Farm Milk Handling Equipment
- DPC 29, Cleaning & Sanitizing in Fluid Milk Processing Plants
- DPC 57, Dairy Plant Sanitation
- DPC 91, Conducting and Documenting HACCP, SSOPs and Prerequisites
- DPC 92, Conducting and Documenting HACCP—Principle Number One: Hazard Analysis
- DPC 93, Conducting and Documenting HAACP—Principles #2 and 3, Critical Control Points and Critical Limits
- DPC 94, Conducting and Documenting HAACP—Principles #4 and 5, Monitoring and Corrective Action

mind that the creation of a great plan takes time. Creating a complete risk reduction plan is a process; if you want to learn advanced math, you don't start with calculus. Start by creating a foundational program that addresses each step in the plan in a basic way; then whenever possible make each step more comprehensive. It has taken us five years to get to where we are, and thanks to more classes and training, every year we learn more. Keep your plan in a binder dedicated to the topic, with separate sections for each step, along with anything else related to your plan.

The seven steps are these:

- Adopt good manufacturing practices.
- Write step-by-step cleaning, sanitizing, and operating procedures.
- Make product and process flow charts.
- Create the main policy and procedure documents.
- Develop monitoring records.
- Create the HACCP part of the plan.
- Create a written recall plan.

Again, take this in little steps, always knowing that something is better than nothing! Appendix B contains several sample forms and charts. Let's get started.

1. Adopt good manufacturing practices: GMPs are standards that are required by law for all food manufacturing companies. GMPs are the backbone of each procedure in the risk reduction plan. They are a part of the Federal Code of Regulations, Section 21, Part 110, and can be found online. They also can be found in the Dairy Practices Council (DPC) guidelines, numbers 8 and 91. Understanding GMPs is the foundation for creating good policies and the steps needed to ensure that those policies are followed, so by all means spend a little time reviewing current GMPs, either on the FDA's website (which is not very user friendly) or through the DPC booklets. Since these standards are written by the regulators, feel free simply to copy them and include that copy in your plan. GMPs cover the following four areas:

- **Personnel hygiene:** Everything to do with people, including policies on health, handwashing, attire, jewelry, eating and chewing gum, and cleanliness.
- **Building and facilities:** Everything to do with the building and space, including policies on handwashing facilities, lighting and ventilation, storage of ingredients, separation of finished product from unfinished, pest control, and construction and maintenance standards.
- **Equipment and utensils:** Policies regarding the quality of equipment and utensils—such as ladles, milking equipment, and milk cans.
- **Production and process controls:** This section covers policies on such things as records for ingredients, production batch and inventory records, and temperature control records.

2. Write step-by-step cleaning, sanitizing, and operating procedures: Standard sanitation operating procedures (SSOPs) are written instructions for how to clean and sanitize anything related to supporting the above policies. Standard operating procedures (SOPs) are almost the same, but instead of step-by-step instructions for cleaning, they tell how to operate a piece of equipment or tool properly; for example, how to calibrate a thermometer used to monitor milk temperatures. SSOPs and SOPs should be so well written that an untrained person could pull one out, follow it, and complete the task successfully—ideally. Although it would be lovely to have someone come in and write up all of these policies at once, it is not likely feasible for most small producers—nor is it necessarily going to help you thoroughly understand the plan. So start with one task at a time, and write how-to instructions as if you were writing them for someone who is going to farm-sit. These instructions should tell people what to do if something isn't right; for instance, if an inflation falls off the cow during milking and lands on the parlor floor. You even can do this on index cards and keep it in a little file box. Here is a list of some of the important SSOPs and SOPs for the small dairy producer:

- **Milking equipment setup**: Include inspecting for cleanliness, presanitation, and how to assemble.
- **Animal milking prep:** Include observing animal health and cleanliness, observing foremilk, and udder cleaning protocols.
- **Animal milking procedure:** How to use the milking equipment and monitor proper functions, such as vacuum pressure and milk flow.
- **Milking equipment cleanup:** Include time, temperature, and proper chemical dilutions.
- **Milk filtering:** Include inspecting filters for debris and how filters should be stored.
- **Milk chilling instructions:** How to operate chilling equipment.
- **Milk chilling equipment:** Cleaning, sanitation, and maintenance.
- **Receiving milk jugs:** Include instructions for inspecting new shipments, documenting lot numbers, and storage.

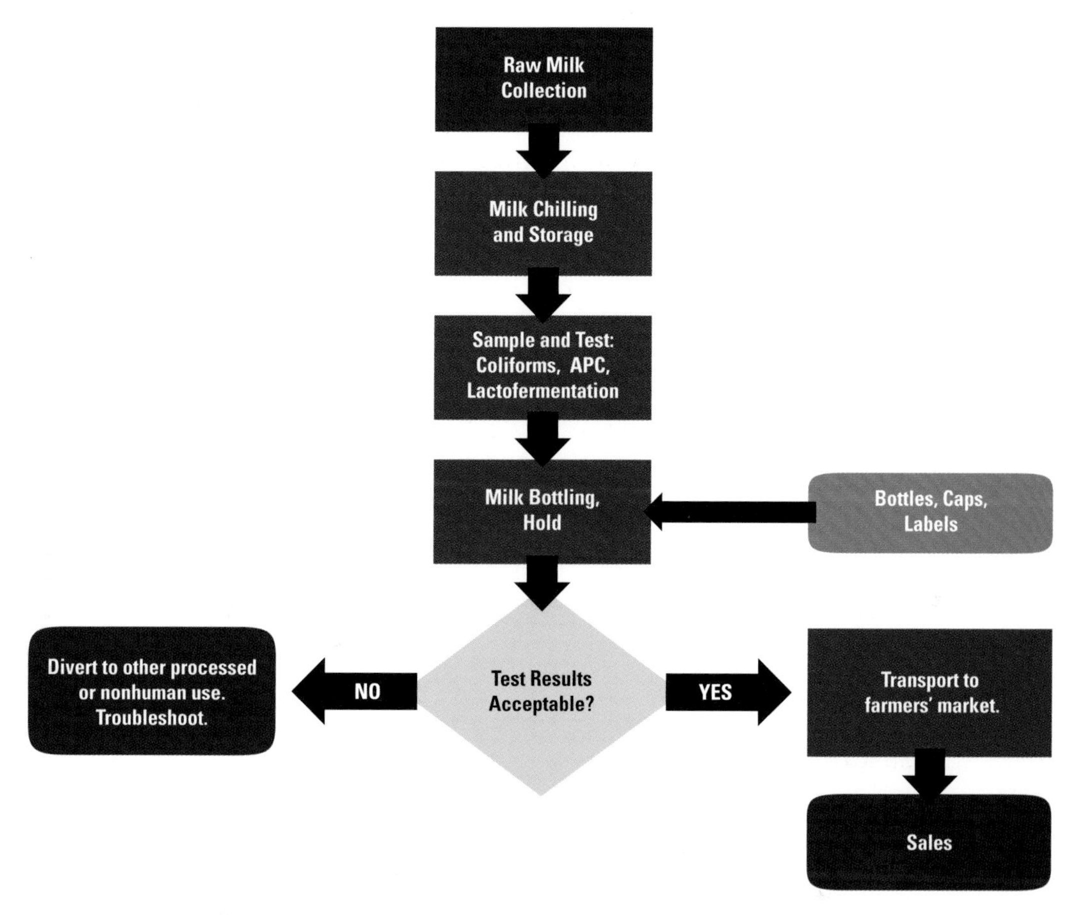

- **Milk bottling equipment:** Inspecting, cleaning and sanitizing, and maintenance instructions.
- **Refrigerators:** Cleaning and sanitizing instructions.
- **Thermometers:** Inspection and calibration instructions.
- **Milk bottling:** Detailed instructions on how milk bottling is to be done, including labeling and tracking batches and microbiological testing data.

3. Make product and process flow charts: Write a description of your product and make a flow chart showing every step of the process and any other products that are brought into the process, such as bottles and caps, from milk collection to sales. These charts or chart will give you a big-picture view of your process and allow you to see all the steps and issues that need to be addressed in each individual policy and procedure (P&P) and will also be used during step 6 (see sample).

4. Create the main policy and procedure documents: These are policies based on accepted standards for the safe manufacture of a food product. This section of the plan probably will take the most time. In the sidebar I have listed thirteen policies that must be included, at minimum; your facility and process may require even more, especially if you have other activities going on in the same place. In appendix B there is a blank sample policy and a completed sample that should give you

MAIN POLICIES FOR THE RAW-MILK PRODUCER

- **Water:** Addresses the microbiological safety of all water that comes into contact with any surface that has direct contact with milk, from milking hoses and pails to milk bottles. The policy should state the source of your water (such as a private well or a municipal source), your water testing policy, and any preventive programs that you might have in place (such as antibackflow devices that prevent livestock tanks from siphoning through a potable system, chlorination, or UV filtration).
- **Product contact surfaces:** Address the condition and cleanliness of any surface that could have contact with the product, such as hoses, milk pails, ladles, and milk containers. Condition includes such things as food grade, cleanable, and in good repair. (See sample Policy and Procedure in appendix B.)
- **Environmental cleaning and maintenance:** States your policy and procedures for cleaning, sanitizing, and maintaining the dairy processing environment, including refrigerators, coolers, and so forth.
- **Prevention of adulteration of product:** Addresses the prevention of contamination of all contact surfaces, utensils, gloves, packaging, and so on, by chemical hazards, such as antibiotic residues, cleaning compounds, and lubricants; physical hazards, such as bandages, fingernails, jewelry, and pests; and biological hazards, including bacteria and allergens.
- **Maintenance of handwashing and toilet facilities:** Explains how you will ensure that handwashing sinks and toilet facilities will be available and fully functional for all dairy personnel. Things to include are the availability of a good hot water supply; clean, single-use towels, and soap.
- **Correct labeling and storage of all chemicals:** A short but critical policy that states your awareness of chemical hazards through improper labeling and storage. Also should mention how each chemical is properly diluted and used.
- **Control of worker health conditions:** Puts in writing your policies regarding sending people home who are ill and excluding them from certain tasks when contamination of product from activities such as coughing and sneezing could occur.
- **Exclusion of pests and pets from dairy and processing areas:** Addresses why and how pests and pets will be kept from the dairy.
- **Receiving of ingredients and packaging materials:** List your protocol for documenting items such as lot numbers on ingredients and packing materials that come into the process. Tells how these materials will be stored to prevent contamination after you receive them.
- **Temperature controls:** Addresses how and why you will monitor milk temperature throughout every stage of the process. Includes that thermometers are properly calibrated.
- **Inventory and traceability:** States how and why you will document each batch of milk so that it could be recalled efficiently if necessary.
- **Worker training:** Even a tiny family farm should document how each worker in the process will be trained in the farm's risk reduction plan.
- **Visitor policies:** This policy should include that visitors must read and sign a sheet, or log, that lists your policies for visitors being in the facility, such as health status, jewelry, hairnets, and clothing. Every person visiting, including regulators, should sign this policy, and the date and times of their visits should be logged.

a good idea of how these need to be written and how each one will be integrated with step 5, monitoring records. For now, just start out by printing a template for each of the thirteen policies and start filling them in for each of the next four areas:

- **Write the policy goal:** The goal of each policy should be based on official GMPs. These policies cover personnel hygiene, building and facilities, equipment and utensils, and production and process control. You can base these on the descriptions in the sidebar.
- **Write the procedure:** The steps in each procedure will describe how the goal will be achieved. These steps incorporate SSOPs and SOPs. SSOPs define the proper cleaning and sanitation of equipment and facilities. SOPs define how to do something, such as calibrate a thermometer, receive a new shipment of milk jugs, and document your milk sales. Instructions for cleaning, sanitizing, and operating can be kept in a binder in the appropriate room so anyone needing to do these tasks can access the instructions easily.
- **Define how the policy is monitored:** Each policy should define how it will be monitored. A monitoring log or checklist (covered in the next step) will be a separate document that is referred to in the policy. There is a saying, "If it isn't written down, it didn't happen." I'll cover some fairly user-friendly monitoring options in a bit, but for now make sure you understand that the policy should say how it will be monitored. Take a look at the sample policy in appendix B, and note the monitoring logs that are referred to. Also you can look at the sample monitoring log and see the policy that it covers.
- **Define how to correct problems:** The policy should have a statement that simply sums up what you will do to address problems and deviations. This is often an identical statement for each policy, so it might seem redundant to most of us. But it is an important feature that helps satisfy officials.

5. Develop monitoring records: One of the trickiest parts of the risk reduction plan, especially for the small multitasking milk producer, is creating a monitoring system that is user friendly enough that it actually can be completed instead of avoided. On the other hand, the small producer is often his or her own supervisor, so oversight is much simpler. For the small producer, checklists can be consolidated so that several policies are covered. For example, you might have a checklist that is called "Milk Bottling Room Checklist."

At our farm there are only two of us in the cheese room at any one time (unless we are holding a class). We use check marks instead of initials. I am right-handed, and Vern is a lefty, so his checkmark goes in the opposite direction (I call that backwards) from mine. Good enough! Be sure to keep a blank copy of each record in your risk reduction plan binder, then keep a separate binder with archived records. There is a sample monitoring log and process log in appendix B.

6. Create the HACCP part of the plan: Finally, we get to the part that these plans are named after! This consists of three steps that are closely linked but that should be documented separately. Appendix B includes a sample of both the hazard analysis and critical control point forms

a. **Do a hazard analysis:** Create a table showing each step of the process (using the flow chart). Include any materials or ingredients that enter the process. Each step of your process is analyzed for any hazard. The three areas that hazards fall under are:

   - **Chemical:** Examples include antibiotic residues from animals that have been treated; sanitizer residue from improper use of chemicals; pesticide from poor pest management; paint from poorly timed maintenance; vinyl bits from gloves (as an allergen); and allergens from other products also being processed.
   - **Biological:** Examples include growth of or contamination by pathogenic bacteria that might enter the milk at several points during the process; viruses, such as hepatitis A and norovirus that might enter the milk from workers; and bacterial contamination from humans.
   - **Physical:** Examples include bandages, fingernails, disposable glove fragments, glass, and hairs.

b. **Identify critical and noncritical controls:** A step that must be done at a certain point during the process to prevent, limit, or eliminate a food safety hazard is called a control point. If the hazard could be life threatening, the point is a critical control point (CCP). For example, in conventional milk plants pasteurizing the milk is the critical control meant to eliminate bad bacteria that could be in the raw milk. After heat treatment other controls are implemented to prevent postcontamination of the milk.

   As you look at your flow chart or table and the hazards that you have listed, you must decide which are control points that are dealt with by your policy and procedure (P&P) and which are critical control points. Although the CCPs might already be addressed in your P&Ps, by listing them and readdressing them in this separate hazard analysis, you are documenting awareness of the most dangerous aspects in your process. So it may seem a bit like extra work, but the hazard analysis is a brief overview that highlights the risks, an overview that officials can look at quickly without having to sift through the many other documents in your plan.

c. **Identify and define specific critical control points:** This step may seem like yet another duplication of your hard work, but if you look it over closely, you will see how it is a great partner for the table above. Combining them in one table would be confusing! You probably will hear some arguments that there is no such thing as a reliable CCP for fluid raw milk because pasteurization is seen as the best CCP to reduce or eliminate pathogens. But since a CCP is defined as a step that can prevent, eliminate, or reduce a risk, I disagree. True, the degree to which the risk is reduced will be somewhat subjective and unpredictable. Labeling that identifies raw milk as a food in a higher risk category is completely acceptable and advisable. Producers should not feel that this is a personal insult to their product; it simply points out a generally accepted categorization and is meant to place the burden of choice on the consumer. Appendix B contains a sample CCP chart.

7. Create a written recall plan: The FDA has the power to order a recall of your product. When I wrote my first book, no recall was mandatory—in theory. A recall might be for as innocent a reason as putting the wrong label on a product. Not all recalls are death sentences for the producer, but even a low-risk recall can become a nightmare if the producer does not handle it well. A recall plan will help you become a "political thinker" during a delicate situation.

Most small producers who sell all their milk directly to the consumer from the farm, through distribution via a herdshare, or delivery or sales at farmers' markets, will find that an extensive third-party certified recall plan is both unnecessary and impossible to create. If there is no middleman, a recall plan can be abbreviated. No matter what size processor you are, please still create a plan. You will be amazed at the information that—heaven forbid, you should ever need it—will be right at your fingertips during an impossibly stressful time. Here are some things to include in a good recall plan:

- A list of customers. Usually these are retail customers. If you sell at a farmers' market, simply list the market manager's contact information.
- A list of officials and media along with their contact information. Create this list, and plan to make your calls in order of priority.
- The FDA regional contact if the recall is instigated by you
- The state-level regulators
- County and local regulators (department of health or its equivalent)
- Any organization related to your industry that might need to know
- Local press
- Describe how you track each batch of milk—through date, lot number, and labeling.
- Have on hand some templates for press releases regarding various typical recalls. See the bibliography for a link to the CDC's recall templates.
- Create "talking points" that you will stick to if you are contacted by the press. Talking points can be created using the templates in the previous step. I can't emphasize enough the importance of having these. If you ever do experience a recall, you will be extremely challenged to avoid emotion, defensiveness, and statements that will come back to haunt you.

PART IV

MOVING BEYOND THE BOTTLE

·12·

Fermented Milks: Ancient Foods for Modern Times

There is little doubt that fermented milks were among the first processed foods consumed by humans. Anne Mendelson in her book *Milk: The Surprising Story of Milk Through the Ages* goes so far as to refer to ancient eastern Mediterranean and Fertile Crescent as "Yogurtistan." Early humans no doubt did not trade recipes and purposefully make cultured milks, rather they collected milk and fermentation occurred spontaneously. Over time, traditions related to fermentation arose and developed. Although some of the names of the fermented milks widely available today barely resemble those of their motherlands, they are all still fantastic ways to extend the life of milk and increase the health benefits—and, as a consequence, also reduce risks. And of course, make delicious foods!

All of the following recipes can be made from the high quality raw milk of cows, goats, or sheep. By doing so, you will end up with products unlike any of those available commercially—unless you live in a region in which they are available made from raw milk. Yogurt can, in theory, be made from raw milk, but it will be quite a bit different in both flavor and texture than traditional yogurt. Fermented milks are just one step away from cheese. For the purpose of this book, I define cheese as cultured or acidified milk that also has been concentrated through drainage—of the watery part of milk. People do, in fact, drain yogurt to the point at which it is called yogurt cheese. If you are interested in taking the next step in cultured milk products, raw or otherwise, my last book *Mastering Artisan Cheesemaking* is available to help.

Remember that cultured, fermented milks have the added health benefits of beneficial bacteria that are added to or already exist in the raw milk and which will have been encouraged to flourish. Depending upon the amount of acid produced by these bacteria, the milk will have become a less hospitable place for spoilage or pathogenic bacteria. In addition, the presence of acid means that bacteria have consumed much of the milk sugar and have produced some lactase. This can mean that people with mild intolerance to lactose might be able to digest these types of fermented milks more easily. If you are looking for an extensive collection of fermented milk recipes, along with fascinating history and perspectives, I recommend buying a copy of Anne Mendelson's book (mentioned earlier) as well as *The Art of Fermentation* by the King of Fermentors, Sandor Ellix Katz.

A TALE OF TWO CREAMS—BOVINE AND CAPRINE

If you have access to fresh cow's milk, then you know the beauty of bovine milk's ability to "cream." Large fat globules easily clump and rise in cow's milk, thanks in part to the presence of a protein type called "cryoglobulin" that attracts fat globules. Think of the fat globules as little balloons rising to the top of the milk. As the cryoglobulin collects the little balloons, they form large groups of balloons that rise even more quickly. Goat's milk not only has smaller, less dense milk fat globules, but it also lacks the clumping protein. Goat's milk eventually will separate, but it takes much longer and never will be as complete as what happens with cow's milk.

If you have goats, particularly those that produce high fat milk such as Nubians, Nigerians, Pygmy, and even Boer goats (yes, you can milk a meat goat female), it is quite possible to collect cream—without the use of a mechanical separator. All that is required is patience. First, chill the milk in a wide mouth jar (the shallower and wider the container, the faster the cream will separate). Once the milk is chilled to full refrigeration temperatures, move the container to the freezer and let it sit for an hour or two. Don't let it freeze; just allow it to get closer to freezing. This will encourage the cream to rise to the top. Then move the container back to the refrigerator and let it set for a day or two. Once a layer of cream has formed, skim it off carefully and place it in a freezable container. Place in the freezer and keep adding more cream to it until you have enough to make your butter or crème fraiche. Remember that goat's milk lacks the yellow tinting pre-vitamin Beta Carotene. So brace yourself for some snowy white products—that may look a bit reminiscent of Crisco shortening, but will taste fantastic!

Just churned butter being pressed and worked to remove true buttermilk.

Cultured Butter and True Buttermilk

If you are lucky enough to have access to farm-fresh raw cream, you are in for a treat when you make cultured butter. Most commercially available butters are made without the addition of any starter cultures, although some brands such as Organic Valley and Vermont Creamery make lovely pasteurized cultured butters. Butters can be cultured naturally by allowing the raw cream to ripen for a period of time—allowing the natural bacteria to develop—or through the addition of ripening starter cultures, or both! Raw-milk butters can be made with or without culturing. Cultured butters have a more complex flavor and aroma, thanks to the development of a small amount of acid, as well as flavor and aroma compounds produced by the bacteria, but they are also more prone to rancidity.

After the butter has been churned and pressed, the leftover liquid, with its tiny flecks of golden butter (assuming you are using milk from grassfed cows) is true buttermilk. True buttermilk can be a beverage on its own or used in cooking. You also can make goat's and sheep's milk butters, provided you use a cream separator to spin the cream from the milk or if enough cream can be skimmed and collected over several days. (When collecting enough cream over a period of days, I recommend storing it in the freezer until you are ready to make butter.) For high-fat goat's milk, such as from Nigerian Dwarf and Nubian goats, skimming cream is a possibility.

Ingredients

Fresh or frozen raw or gently heat-treated cream
Starter culture such as Flora Danica, buttermilk, or crème fraîche (these types usually contain aroma cultures) or cultured buttermilk with active cultures
Pure salt (without any additives)

Equipment

Sanitized jar and/or butter churn
Bowl
Fine-mesh strainer or butter muslin (fine cheesecloth)
Spoons or butter paddles (also called Scotch hands)

Procedure

1. Collect the cream to be churned. If you are collecting from more than a few days' worth of milk, you can store the cream in the freezer until you are ready to make butter.
2. Bring the cream to room temperature (about 70°F [21°C]), and add cultures. Use a scant ⅛ teaspoon (about 1.75 gm) of freeze-dried culture or ⅛ cup (30 ml) of cultured buttermilk for each quart (or liter) of cream. Allow the cream to sit at room temperature (ripen) for 12 to 24 hours.
3. Cool the cultured cream to about 55°F (13°C).

4. Pour the cream into the churn, or leave it in the collection jar, as long as about one-third of the jar is empty.
5. Churn or agitate the cream until flecks of butter are visible. This may take 5 to 10 minutes. If you are using a jar, simply shake the jar up and down or back and forth. When the fat globules begin to cluster, you will hear a distinctive change in the sound and feel and will see the glob of butter forming. Once the butter flecks appear, stop churning.
6. Drain the buttermilk off, either using a fine sieve or a piece of butter muslin (fine cheesecloth).
7. Rinse the collected butter bits with cold water. This helps remove more buttermilk and encourages the fat to firm up.
8. Transfer the butter to a cool bowl, and gently press and work it into a ball. You will see buttermilk ooze from the mass. Using cold spoons or paddles will help prevent the butter from sticking to them. You can keep two sets going, with one soaking in ice water while you work with the other set.
9. When you like your butter's texture and dryness, work in a tiny bit of salt to taste. The amount of salt you add is completely up to you.
10. Press the butter into a tub or form, chill completely, and use. It can be frozen to extend its shelf life.

Troubleshooting

- If butter doesn't form, but you end up with whipped cream, the cream was too cold.
- If butter forms but it is too whipped in texture and doesn't separate well, the cream was too warm.
- If the butter is greasy and soft, you churned it for too long.

INCUBATION OPTIONS

When you ripen fermented milks, it is important to maintain the temperature as close to the goal of the recipe as possible. You may have a warm place in your home that will work well, but if you don't, one option is to use an ice chest to keep the milk at a steady temperature. After adding the cultures, fill a jug or other closed container with warm water, at about 2°F (1°C) warmer than the ripening temperature. Put the milk and warm-water bottles in the ice chest, then pack a few bath towels in around it all. Close the lid, and wait the set amount of time.

A glass of raw, cultured buttermilk.

Cultured Buttermilk

When I was a little girl, my mother used to prescribe buttermilk for upset tummies (she included a sprinkling of black pepper on top). I can still picture the partly empty glass with the characteristic traces left by the bits of butter and pepper coating the sides. I still associate this flavor combination with the wonderful, slightly paradoxical feeling of being ill but lovingly tended. Cultured buttermilk is often overlooked as a delicious beverage and is more often used for baking biscuits and making salad dressings. Buttermilk might be one of the best choices you could make, however, if you want to turn raw milk quickly into even more of a superfood for your family. You can flavor it with fruits, natural sweeteners, and spices—both savory and sweet—to encourage kids to develop a love for this overlooked wonder.

Ingredients

Fresh raw or gently heat-treated milk
Starter culture such as Flora Danica or cultured buttermilk from a previous batch or the grocery store (as long as "live active cultures" are listed as present)

Equipment

Sanitized jar
Thermometer

Procedures

1. Bring milk to room temperature, about 72°F (22°C) (you can use it straight from the milking parlor if you like). Pour it into a sanitized jar.
2. Stir in about ⅛ teaspoon (2gm) starter culture or ⅛ cup (30 ml) buttermilk per quart or liter of milk.
3. Maintain the temperature (ripen) for 8 to 12 hours.
4. Chill the buttermilk, then taste it. You can adjust the tartness of future batches by shortening the ripening or lowering the ripening temperature just a bit (for sweeter buttermilk) or ripening for longer or at a warmer temperature (for more tart buttermilk).

Troubleshooting

- If the buttermilk is too tart or sour, lower the temperature during the 8-to-12-hour ripening time or decrease the ripening time.
- If it is too thin or too sweet, increase the ripening time or the temperature.
- If there are air bubbles or the buttermilk is frothy, the milk is contaminated with yeasts or coliforms from contamination during milking. Throw it out. If this problem occurs frequently, you need to address milking hygiene. Consider heat-treating the milk for safety's sake and the quality of your products. Note: The starter culture Flora Danica can produce some effervescence, but I have not seen that happen if the ripening occurs within the time suggested.
- If it lacks flavor, try a different culture. My favorite is Flora Danica.

SOURING NATURALLY

Most of these fermented milks, with the exception of yogurt and kefir, can be made without adding starter cultures if you have enough good lactic acid-producing bacteria in the milk. In chapter 10, I covered how to do lacto-fermentation tests to get a good idea of what kind of bacteria you have in your milk. If you have done these tests or are simply adventurous, you can try ripening milk for butter, buttermilk, and crème fraîche without adding cultures. If you want the best chance of a more consistent, flavorful, and safe product, however, you should add a bit of purchased cultures.

Kefir

A big thank-you to Alethea Swift from Fiore di Capra (see her sidebar profile in chapter 10) for providing this kefir recipe. Fiore di Capra is a licensed raw-milk and cheese dairy in Arizona.

Kefir is a milk product fermented by a symbiotic colony of bacteria and yeasts (SCOBY), a paradoxically repulsive yet beautiful gooey mass. Several other products are fermented by these "creatures," including vinegar, where it is called the "mother"; a fermented tea called kombucha—where it is called a "mushroom" or mother; water kefir, where it is called tibicos; and a Tibetan cultured milk called tara. Kefir SCOBYs also are called "grains" because of the way the mass grows in little grain-size nodules (see photo). Each SCOBY contains microbes that prefer to utilize the nutrients present in the solution they ferment. Kefir grains readily ferment milk sugar and in the process produce lactic acid, beneficial bacteria and yeasts, and the tiniest bit of alcohol, leading to kefir's nickname, "the champagne of milks."

Ingredients

Fresh raw or gently heat-treated milk

Kefir grains, either fresh or dehydrated. You will need about 2 tablespoons (30 ml) of fresh grains per 2½ cups (0.6 L) of milk. If you are using dehydrated grains, follow the package instructions.

Equipment

Clean glass mason-type jar sized to fit the amount of milk you intend to culture. Allow for space at the top.

Spoon

Thermometer

Nylon mesh or stainless steel strainer

Procedures

1. Be sure you start with a very clean glass jar and lid. Kefir works best between 65 and 85°F (18 and 29°C). If you use milk that is fresh from the animal, be sure to cool it a bit before adding your grains. If you are using prechilled milk, warm it up a bit first. If your room temperature is below 70°F (21°C), it is recommended to start the kefir process at the higher end of the range, about 85°F.
2. Place your kefir grains in the jar, and add the milk. Do not fill the jar more than three-quarters full, since the kefir will produce carbon dioxide. Place the lid on the jar, and gently swirl the grains around in the milk. Now loosen the lid slightly to allow gas to escape during the fermentation process. Let the jar sit at room temperature for 24 hours. Do not place the jar in direct sunlight.

Healthy clumps of live kefir "grains."

3. Stir and test the kefir after 24 hours. It will be sour, thickened milk at this point. If you prefer a more sour kefir, you can let it ferment for 6 to 12 more hours. Check it every 3 to 4 hours with a clean spoon. Once the kefir is at the desired consistency, strain it into another clean jar. If you are using fresh grains, you will need to start another batch of kefir. Be sure to clean the fermentation jar before you start a new batch, or use a fresh jar. The grains do not need to be rinsed between batches, but if you want to rinse them, do so in raw milk rather than water. The fresh grains will grow over time, and they can be separated and shared with a friend.

Troubleshooting

A word about kefir cultures: One advantage to using dehydrated grains is that you do not have to make a new batch of kefir every 24 hours. Just store the packets of dehydrated grains in the freezer and use them whenever you want to make up a batch of kefir.

Variations

- **Honey-Vanilla Kefir:** This is a favorite among our customers and our family. Add about ¼ cup (60 ml) of raw honey and ¼ teaspoon (1.2 ml) of vanilla bean paste (or extract) per quart of kefir. Warm the honey first for easy mixing. You can pour this Honey-Vanilla Kefir into ice pop molds to makes delicious, healthy ice pops.
- **Kefir Cheese:** Allow your kefir to sit at room temperature for approximately 48 hours. Strain it through butter muslin or tightly woven cheesecloth until it stops dripping. You will have a soft cheese similar to cream cheese.

SCOBY SCOBY DO! WHAT ARE YOU?

You can almost think of a SCOBY as the beginning of a new life-form. Our own human cells are thought to have evolved when microbial organisms formed symbiotic communities for mutual benefit. This theory, called endosymbiosis, was advanced and substantiated by the pioneering biologist Lynn Margulis. The microbes, bacteria, and yeasts in a SCOBY don't simply ferment sugars; they manufacture the structure that binds them together *and* they grow and reproduce. According to Sandor Katz in his book *The Art of Fermentation*, Dr. Margulis used kefir grains as an example of how cells can organize, work together, and create a new and visible form. These life-forms not only contain many unknown and unnamed microbes, but they do not have what scientists call a "programmed death" such as do humans, trees, and most of the things that we think of as alive. In fact, bacteria are the oldest living organisms on the planet.

Homemade yogurt that is naturally thick thanks to high protein, butterfat, and calcium content.

Yogurt

Yogurt is the granddaddy of fermented milks. Its importance to early civilization is so great that, as I mentioned in the introduction, author Anne Mendelson even dubbed the geographical regions in which yogurt originated "Yogurtistan." If you have already made the cultured buttermilk recipe earlier, you will notice many similarities between it and yogurt. As with most cultured dairy products, it is often little things that end up making a big difference in the end product.

Unlike buttermilk and many cheeses, yogurt ferments with the help of thermophilic, or heat-loving, bacteria. So it is important that the ripening of yogurt take place at temperatures ranging from 110 to 125°F (43–52°C), with 110°F being the most common. The temperature will influence the tartness; more on that in a moment. Bacteria for making yogurt usually contain similar strains of the same type of bacteria, with a little variation, and different ratios. You might want to experiment with several, as well as some of the "heirloom" cultures now available from such companies as Cultures for Health (listed in the resources). Not only will flavor vary, but so will tartness and texture.

Some cultures will work better with some milk types than others. The incubation (ripening) temperature range can be used to control the balance of sweet to tart, with the cooler temperature producing less acid and tartness. But the direction you nudge the temperature in and the change that results will vary depending on the culture! In other words, purchase a lot of different starters if you can, and keep track of your experiments.

Many commercially produced yogurts include some kind of thickener, whether that be dry milk solids, pectin, or other ingredient added to prevent whey from separating from the coagulum. For home use there is no reason to add anything to your yogurt. If you like it thicker, drain it. If it is thin and you don't have time to drain it, you have made drinkable yogurt. Now that Greek-style yogurts are popular, more commercial companies are draining their products to obtain the desired texture—but the cost is greater and the yield of product less.

Ingredients

Fresh, raw milk. It can be whole or skimmed.
Plain unflavored yogurt with active cultures (can be from previous batch)

or freeze-dried yogurt culture. (You can keep reusing the same yogurt for an indeterminate amount of time; the batches will vary, and you will have to decide when it is time to update to a new culture.) Use 2 tablespoons (60 ml) of yogurt or ⅛ teaspoon (0.4 gm) of freeze-dried culture (or as recommended on the packet) for every quart of milk.

Equipment

Heavy saucepan
Spoon and whisk
Thermometer
Sanitized glass jar and lid
Fine cheesecloth (often called butter muslin) and colander if yogurt is to be drained

Procedures

1. Pour the milk into the pan and place on direct heat (or use a water bath if you don't want to stir constantly). Stir and heat the milk to 185 to 190°F (85–88°C). You can increase the temperature to just below boiling if you want an even thicker, higher-protein yogurt. The high temperature denatures proteins that would otherwise want to separate in whey, helping create a thicker body.
2. Remove the milk from the heat and set in a sink of cold water. Stir until it cools to between 115 and 125°F (46–52°C).
3. Place the yogurt culture in a bowl, and add about 1 cup (250 ml) of the warm milk. Whisk to blend. If you are using freeze-dried cultures, sprinkle them on top of the warm milk in your bowl, then whisk to blend. Add combined milk and cultures to the pan, and whisk well.
4. Pour into jar, cover, and place in a small ice chest. You can tuck a towel around it to help keep the heat in. Set where the room isn't too warm or cold, and allow it to incubate for 4 to 8 hours. Check the consistency and temperature at 4 hours. If it's not set enough, incubate longer.
5. When the right texture and flavor are achieved, place the jar in cold water or directly in the refrigerator. As it cools, the yogurt will thicken more.

Variations

- **Greek-Style Yogurt:** Once the yogurt has set, stir gently, then pour into a cloth-lined colander that's set in a saucepan or large bowl. The colander should sit up away from the bottom of the pan or bowl by several inches. Cover, and let it drain to the desired texture. The draining area should be cool but not cold. You may need to stir it occasionally during draining. Usually 2 to 4 hours is plenty. Drained yogurt will last longer and not grow more tart as quickly during storage. Chill thoroughly after draining.

- **Yogurt Cheese:** As with the kefir recipe earlier, yogurt can be drained to a thick paste, suitable for spreading, much in the same way as cream cheese. Continue as for Greek-style above, but let it drain even longer. A touch of salt will help speed draining and help keep the yogurt from growing more tart.
- **"Rawgurt":** If you want to make a yogurt-like cultured product and keep the milk raw, you can! Out of respect for the long history of yogurt, I would encourage you to not call it by the same name. Heat the milk to only 115 to 120°F (46–49°C), then follow steps 3 through 5. This method is likely to make a much thinner, drinkable product, but it will still be delicious. It won't have quite the number of probiotic bacteria as the original, because of the competitive growth of the bacteria already in the milk (some of which may be probiotic). This type of cultured milk is almost like a buttermilk-yogurt hybrid.

Decadent, versatile crème fraîche

Crème Fraîche, Sour Cream, and Kin

Although the French translation means "fresh cream," crème fraîche is actually a soured or cultured cream that originally would have been made with fresh, raw cream naturally ripened to a thick, tangy, spoonable loveliness. Devonshire clotted cream, American sour cream, Italian mascarpone, Mexican crema, and even cultured butter are all relatives of crème fraîche (but don't tell this to the French). Instead of thinking of sour cream and crème fraîche as two different recipes, think of the end product as a cultured cream of varying thicknesses and tang, depending upon the heaviness of the cream, the type of starter culture, and how long it is ripened.

Ingredients

1 pint (2 cups) heavy cream. Choose the purest, freshest possible pasteurized cream. Most commercially available cream has added thickeners, stabilizers, and sweeteners. Many are also ultrapasteurized. If you have access only to those with a few additives (that's what I made mine from), you still can make great crème fraîche. But don't try to use ultrapasteurized cream; it will not thicken properly.

Culture. Choose one of the following:

- 3 Tablespoons (90 ml) buttermilk with active cultures (it should be labeled as such)
- 3 Tablespoons (90 ml) yogurt with active cultures (slightly different directions in step 4)
- 3 Tablespoons (90 ml) sour cream with active cultures (hard to find)
- 1 packet direct-set crème fraîche starter
- ¼ teaspoon (0.8gm) Flora Danica starter culture

Equipment

Pint jar with lid

Steps for Making Crème Fraîche

1. Sanitize the pint jar and lid with boiling water. (The jar must have been manufactured for use in canning or it might crack when you pour the boiling water onto it.)
2. Warm the cream in your stainless steel saucepan on the stove, stirring constantly, to 86°F (30°C).
3. Remove your cream from the heat, and stir in your culture of choice until it's evenly mixed.
4. Pour the mixture into the jar, seal, and let it set at about 72°F (22°C) for 12 hours. (If you are using yogurt as a culture, incubate at a warmer temperature of 85 to 95°F (29–35°C)). If needed, you can set the jar in a small ice chest with another jar filled with warm water to help maintain the temperature.

5. After 12 hours check the consistency and flavor. If it's ready it should have a gravylike consistency and a tangy taste. The pH (if you can check it) will be about 4.55.
6. If you are satisfied with the flavor and texture (don't worry; it will get thicker), place the jar in the refrigerator for 24 hours. During this time it will thicken to a sour cream or yogurt texture.
7. Your crème fraîche is now ready to use! If you desire a bit fluffier consistency, simply pour it into a bowl and whisk to thick peaks. You can sweeten or flavor the crème fraîche as well.
8. Store in your refrigerator. It will keep (even after being whisked) for about 3 weeks (if you used fresh cream), a little less if the cream was stored for long before being used.

If you think that this chapter is just the tip of the iceberg when it comes to the wonderful things you can make from milk, you are right! From puddings and custard to cheese, a bounty of milk can be used to broaden your menu and spoil your taste buds. I have listed a few books that will take you to the next level of dairy product mastery in the bibliography. I recall my mom would make big trays of custard when I was a kid; not only did it use up a lot of milk, but it also made a dent in the eggs that were often overflowing the cartons in the fridge. Looking back, it is funny to find that although we had so little money, in some ways we were quite rich.

· 13 ·

Heat Treatments, Canning, and Freezing

In most cases when farm life follows the seasons of nature, there is a period of time in which milk is not available. Many farmers milk their cows and goats throughout the year (meaning they either stagger the birthings for year-round milk) or they don't rebreed some but instead continue to milk them for more than one season. This provides a supply of fresh milk year-round. But if your farmers and their animals take a break, you are likely to find yourself short of precious raw liquid milk for a number of weeks or more.

Cheese has been the traditional way to store milk. But it is hard to pour on your cereal. If you want to put aside a little liquid gold for the dry season, or if you want to prepare for times of need, then preserving milk through canning or freezing is a great option. Although canning will reduce many of the qualities that you might most appreciate about raw milk, it is a way to set aside nourishment without the need for refrigeration. Freezing milk will change some of its qualities as well but not quite as much as will the heat from canning.

I know those reading this book rarely will need to pasteurize milk, but I want to give you options for the gentlest of heat treatments that can be used on milk just in case. As an example, a number of years ago our goat herd experienced an outbreak of milk-borne illness. A bacteria, called mycoplasma, was transmitted through the milk to a number of baby goats, causing one fatality and several others to be ill and become lifelong carriers of the bacterium. Since there is no definitive test for whether an animal is a carrier and the fact that the bacteria may or may not ever be shed by the carrier, we decided that it was important to heat-treat the milk for the babies. There may be times when that same choice is right for you—even if it is a one-time deal.

Canning Milk

Canned milk has an extremely long life span—several years. Once it's properly canned it does not require any resources, such as refrigeration, to store, making it perfect for families seeking long-term, low-maintenance storage of food supplies.

To can milk properly you need to use a pressure cooker/canner to achieve the right temperatures for ensuring that the spore-forming bacteria that cause botulism are not present. Milk doesn't have enough acid present to help preserve it safely if you can it using the water bath method (the way you can to preserve tomatoes, jellies, and jams). Refrigeration also prevents the growth of this bad bug. The bacteria that cause botulism can enter milk at any time so their presence is not necessarily related to milking hygiene.

Equipment

Pressure cooker/canner
Quart canning jars and lids suitable for canning

Procedures

1. Prepare the jars by first inspecting them for cracks. Then wash them thoroughly and immerse them in hot water (to keep them warm so they don't crack when you pour the hot milk into the jar).
2. Place the lids or rings and lids into a pan of water, and bring it to a boil for a minute or two.
3. Fill the jars to within ½ inch of the top with fresh, filtered raw milk.
4. Wipe the rim, and place a lid on the jar. Tighten to hand tight, so it's snug but not forced to supertight.
5. Place the jars in the pressure cooker, and add the amount of water indicated by the directions included with your canner.
6. Place your filled canner on the stove, and following the manufacturer's instructions, bring to 10 pounds of pressure and process for 25 minutes for quart jars. Maintain the pressure at 10 pounds throughout the processing.
7. Remove the canner from the heat, and let it sit undisturbed until the pressure has reduced to zero.
8. Remove the jars from the canner, and set them on a dry towel. Let them sit undisturbed overnight.
9. Store in a cool, dark pantry or cupboard.

Freezing Milk

As I mentioned at the beginning of the chapter, frozen milk will have a more familiar fresh-milk flavor than canned milk, but it does require storage in a freezer that reaches colder temperatures than that of the ordinary refrigerator's freezer compartment. If it's not stored cold enough, the milk may separate from the fats, leaving a curdled-looking liquid. If you own a "deep freeze" chest or upright freezer, you can store milk well. Milk cannot be frozen in glass because it will shatter, so you have to be okay with using plastic if you want to store milk in this manner.

At our farm we store two chest freezers worth of milk every winter—not for our use but to feed to goat babies in the spring. I like to use freezer-quality ziplock bags. I hate that I cannot recycle them, but those that haven't sprung a leak I can use again for other uses. Milk can also be frozen in plastic milk jugs, but it does take up quite a bit of space in these containers. The sheep's milk industry commonly freezes milk in 5-gallon "dairy bags" for use in cheesemaking, but that size is likely too large for most home uses.

Frozen milk stored in a chest freezer.

Procedures

1. Chill the milk thoroughly.
2. Stir the milk, and pour it into a new freezer bag, leaving about one-eighth of the bag empty. Squeeze out extra air and seal it tightly. I like to set the bag in a tall plastic container to hold it upright while I pour the milk.
3. Lay the bags on a tray in the deep freeze so the bags freeze flat.
4. When you're ready to thaw some milk, simply remove a bag and place it in a tub in the refrigerator until it thaws enough to remove. The tub will catch milk if any leaks have developed.

Heat Treatment: You Never Know When You Might Need It

Heat treatments, sometimes called heat shock or thermization, involve various times and temperatures designed to destroy as many bacteria as possible without changing the other innate, magnificent properties of milk. In the dairy world heat treatment is used on the first milk, colostrum, when it cannot be fed to the babies raw (for example, to prevent passing a milk-borne illnesses from the mother to the offspring. (Colostrum cannot be heated to pasteurization temperatures without turning it into a pudding.) It also is used by cheesemakers as a way to improve milk quality for cheesemaking, mostly in Europe, but some U.S. cheesemakers employ the method as well. Heat treatment is not defined by regulatory limits in the United States, so you will find varying temperatures and times if you research it. Heat treatments such these can be done on the stovetop using a water bath canner or double boiler (so that milk is not scorched). Remember, unless performed in a commercially approved pasteurizer and with regulatory oversight, the milk cannot legally be called "pasteurized." Here are just a couple of possibilities for heat treatment:

- **Colostrum Heat Treatment:** 138 to 140°F (59–60°C) for one hour. Used to heat-treat colostrum, reaching the same effective level of pasteurization but without causing the colostrum to coagulate.
- **Thermization:** 135 to 155°F (57–68°C) for 15 seconds (per *The Food Processing Handbook*, James G. Brennan, ed. [Weinheim, Germany: Wiley-VCH, 2011]). Thermization is used by some as a way to reduce bacteria counts without damaging proteins. Since this is not a recognized and officially sanctioned process in the United States, temperatures used may vary quite a bit. Reduces overall bacteria counts.
- **Vat or Batch Pasteurization:** 145°F (63°C) for 30 minutes. This is the lowest temperature allowed for legal pasteurization. It is believed to kill all pathogens without greatly altering the other constituents of milk.

With all forms of heat treatment, rapid chilling is essential to stop the process and keep the milk from being altered further. You probably can see that there is great variation in the possible temperature ranges and also, therefore, their ability to destroy the maximum number of bacteria present. I cannot advise you on which method will be the right choice for any given situation.

These alternate methods of preserving and preparing a valued food will, I hope, come in handy for you one day. If nothing else, the knowledge of how milk must be treated will enhance your appreciation of this amazing liquid. I don't know about you, but I am grateful to our ancestors who first realized that perhaps the animals they were eating might provide even more if allowed to live. Not only do I love what milk brings to our diets, from cheese and sour cream to butter and ice cream, but I really love the creatures.

AFTERWORD

Raw Milk and Food Rights

This book started out as a short handbook for raw-milk lovers, those who produce it and those who simply pour and enjoy nature's first food. Much like the historical journey that milk itself has experienced, the project went through some rather difficult labor and growing pains, and in the process my own belief in the wisdom of consuming foods as unprocessed as possible has grown and solidified, as has my belief that the right to these food choices must be actively pursued and supported.

The fear of raw milk that health officials propagate and nurture is absurd. The blind fear of microbes is also not only preposterous but is at the root of the current paradigm that breeds the mistrust of raw milk and fuels a growing trend toward immune disorders and frailty. Although many people continue to look to government agencies for protection and to the scientific community for lifestyle guidelines, many of us prefer to take responsibility for our own choices and use common sense and tradition to guide our health.

Food choices need not await the support of data, or what I like to call the "recent studies show" approach to life. Official nutritional guidelines are generated in an attempt to both support health and bolster industry—they are not pure. In fact, these two goals are often mutually exclusive. Scientific research is often funded when an industry's finances will be improved by particular results and products. I see no need for anyone to need to defend their fervent search for good health from those who say, "Show me the research."

Even the venerable Centers for Disease Control and Prevention recently mentioned the proximity of a "postantibiotic era," when the majority of pathogens become resistant to man's overused and abused "miracle" weapon. Meantime, the knowledge of the human biome—that vast population of life-supporting microbes within us, continues to expand and support the wisdom of lifelong exposure to not only whole, minimally processed foods, but also to microbes of all kinds. I imagine a not-too-distant time when the best advice that health officials will give is that everyone, especially pregnant women and little children, frequently perform farm work, eat raw foods, garden or play in the soil, and own pets—all in the name of building a healthy immune system that might be able to fend off these new, drug-resistant bacteria.

My hope is that this book, although not a political treatise on food rights and raw milk, nevertheless will have provided the knowledge that producers need to continue to expand the right to make and sell raw milk and other small-farm products—in the best manner possible. By doing this I hope to play a role in the support not only of food rights, but also small farms and, even more importantly, small farmers—whether they be urban, rural, or simply dreaming. So let's raise a glass (of farm-fresh, unprocessed milk, of course) and toast to our health and our freedom!

APPENDIX A

Resources

Bottles and Packaging

Specialty Bottle, 206-382-1100, www.specialtybottle.com. Wide-mouthed gallon and half-gallon glass jars.
StanPac Dairy Packaging, 979-251-9851, www.stanpacnet.com. Glass milk bottles, large and small orders, custom imprinting.

Cleaning and Sanitizing Supplies

Many of the listings in the "Milking Equipment and Dairy Supplies" section below also carry cleaning and sanitizing supplies for the small producer.
All-QA Products, 800-845-8818, www.allqa.com. Sanitizer test strips.
Ecolab, 800-392-3392, www.ecolab.com. Supplier of bulk chemicals to the food industry. Some great products; you work with a regional representative. Our experience with customer service has been varied.
Nelson-Jameson, 800-826-8302. www.nelsonjameson.com. Many sanitizers, cleaners, and even udder wash and dips. Cleaning brushes, color coded supplies, and more.
Zep, 877-428-9937. www.zep.com. Similar to above listing in volume and variety of products; also varied customer service.

Cultures and Milk Fermenting Supplies

Cultures for Health, 800-962-1959, http://www.culturesforhealth.com
Dairy Connection, 608-242-9030, http://www.dairyconnection.com
New England Cheese Supply, 413-397-2012, http://www.cheesemaking.com
The Beverage People, 800-544-1867, http://www.thebeveragepeople.com
The Cheesemaker, 414-745-5483, http://www.thecheesemaker.com

Laboratory Supplies

Chemworld, 800-658-7716. http://www.chemworld.com. Sanitizer test strips.
Indigo Instruments, 877-746-4764. http://www.indigo.com. Sanitizer test strips, including hydrogen peroxide strips for testing for percent, not ppm.

Micro Essential, 718-338-3618. https://www.microessentiallab.com/. Sanitizer test strips.

Nelson-Jameson, 800-826-8302. www.nelsonjameson.com. 3M Petrifilm, Quick Swabs, Quad plates, InSite listeria swabs, economy incubator, antibiotic residue testing supplies, and much more.

Weber Scientific, 800-328-8378. www.weberscientific.com. ECA Easygel petri plates and solutions, and more.

Land Use and Pasture Management

USDA Natural Resources Conservation Service, http://www.nrcs.usda.gov/wps/portal/nrcs/main/national/landuse/.

Livestock Management

FarmTek, 800-327-6835. www.farmtek.com. Fodder-growing systems, housing solutions, and more.

Kencove Farm Fence, Inc., 800-536-2683. www.kencove.com. Electric fencing options for pasture rotation.

Premier1, 800-282-6631. www.premier1supplies.com. Electric fencing for pasture rotation and more.

Ranch Supply, 855-413-7883. www.ranchsupply.com. Sydell feeders, panels, gates.

Milking Equipment and Dairy Supplies

Bob White Systems, 802-763-2777. www.bobwhitesystems.com. Small dairy equipment and consultation, Vermont.

Caprine Supply, 800-646-7736. www.caprinesupply.com. Goat-milking and home dairy supplies.

Family Milk Cow Dairy Supply Store, 800-306-8937. http://www.familymilkcow.com. Run by Hamby Dairy Supply but easier to search for the small, home producer.

Farm and Ranch Depot, 928-951-8332. www.farmandranchdepot.com. Many barn and dairy supplies.

Hamby Dairy Supply, 800-306-8937. www.hambydairysupply.com. Supplies the small to midsize and up dairy. Very knowledgeable and helpful.

Hoegger Farmyard Supply Company, 800-221-4628. www.hoeggerfarmyard.com. Mostly carries goat supplies but also has a nice selection of home dairy supplies such as butter churns and cream separators.

MicroDairy Designs, 301-824-3689, http://www.microdairydesigns.com. Small dairy milk processing equipment.

Nasco Farm and Ranch, 800-558-9595. www.enasco.com. Dairy and ranch supplies.

Parts Department Dairy Equipment and Supplies, 800-245-8222. www.parts-deptonline.com. Lots of hard-to-find milking supplies, such as SP6000 (a.k.a. IBA) inflations and, not surprisingly, dairy parts.

Milk Quality Testing Labs

Dairy One, 800-496-3344. www.dairyone.com/MilkLaboratories. Branches in New York, Maryland, and Pennsylvania.

The Dairy Authority, LLC, 970-351-8102. www.dairymd.com.

Udder Health Systems, 877-398-1360. www.udderhealth.com. User-friendly milk-quality testing. Several locations in the Pacific Northwest, but samples can be shipped from anywhere.

University of Minnesota Veterinary Diagnostic Laboratory, 800-605-8787. www.vdl.umn.edu/ourservices/udderhealth/home.html.

Organizations and Helpful Sites

Farm to Consumer Foundation, 513-593-9430. www.f2cfnd.org.

American Micro-Dairies, 802-763-2177. www.americanmicrodairies.org.

Farm-to-Consumer Legal Defense Fund, 703-208-3276. www.farmtoconsumer.org.

Weston A. Price Foundation, 202-363-4394. www.westonaprice.org.

Small Dairy. www.smalldairy.info.

APPENDIX B

Sample Charts and Forms

Sample inspection form (forms are available on the Chelsea Green Publishing website at: www.chelseagreen.com/downloads/Inspection.pdf)
Sample policy and procedure form
Blank policy and procedure form
Sample monitoring log
Sample batch log
Sample hazard analysis for milk production
Sample critical control points

SAMPLE POLICY AND PROCEDURE FORM	
Policy #2	***Title: Condition and Cleanliness of Product Contact Surfaces***

Facility Name: Pholia Farm Creamery
Date: January 1, 2013

Policy Goal: Clean and sanitized processing equipment and utensils are essential to the manufacture of safe food products. Processing equipment is cleaned after each day's run and sanitized immediately prior to the next use. Items must be cleaned and sanitized after each use by hand.

Procedures

1. SSOPs for each piece of equipment used.
    a. How to clean cheese vat and drain table
    b. How to manually clean pots, utensils, cutting boards, etc.
2. Preventive maintenance is performed to keep all equipment in good working condition so it can be properly cleaned and sanitized. Equipment product contact surfaces that are worn and unable to be cleaned effectively shall be replaced.
3. Gloves and outer garments that may come into contact with food are kept in sanitary condition.
4. All equipment used in milk processing and cheesemaking meets recommended state and federal standards for food-contact surfaces.

Monitoring Records

1. Make room check list: Current log kept in make room, archived logs kept for three years in binder in tasting room.
2. New equipment invoices and log: Invoices kept in office, log kept in archive binder in tasting room.
3. Scheduled correction log: Main repairs and replacement log that lists and monitors corrections that cannot be done immediately.

Corrective Actions

Corrections will be taken as needed at each step and will be noted on *Make Room Check List.* Any correction that cannot be accomplished immediately will be assessed as to whether the deviation has a potential impact on the ability to produce a safe product. If immediate measures are needed to minimize the effect of the deviation, they will be taken and noted. Corrections that cannot be addressed immediately will be given a time line for corrections in monitoring form *Scheduled Correction Log.*

BLANK POLICY AND PROCEDURE FORM	
Policy #	Title:
Facility Name: Effective Date:	
Policy Goal:	
Procedures	
Monitoring Records	
Corrective Actions	

MILK BOTTLING ROOM CHECKLIST AND MONITORING LOG									
Enter checkmark or value. Monitors policies: 2, 3, 4, 5, 6, 8, 10, and 11									
Task	**Frequency**	M	T	W	Th	F	S	Sn	**Correction**
Bulk tank or milk storage temperature verified (enter temperature)	Daily								
Bottling equipment inspected, cleaned, and sanitized prior to use	Bottling day								
Working surfaces cleaned and sanitized	Bottling day								
Milk samples pulled and quality tests started	Bottling day								
Jugs and caps inspected and properly loaded	Bottling day								
Milk batch log started—temperature, code, quantity Quality tests results	Bottling day and day after								
Chemicals stored and labeled properly	Daily								
Pest control checked	Daily								
Handwashing station—hot water, towels, soap	Daily								
Trash emptied, drains cleaned	Day after bottling								
Thermometers calibrated	Weekly								

SAMPLE BATCH LOG	
Date:	Code:
Milk production date:	Volume:
Start temperature:	Control sample end temperature:
Microbiological testing results:	
APC:	
Coliforms:	
E. coli:	
Other comments:	
Distribution notes:	

SAMPLE HAZARD ANALYSIS FOR RAW MILK PRODUCTION PHOLIA FARM CREAMERY

Process Step	Critical Hazards	Why Are They Hazards?	What Can Be Done to Control Them?	CCP
Raw milk collection	Biological: Pathogenic bacteria Chemical: Antibiotic residue	Pathogenic bacteria and their toxins can cause illness, and antibiotic residue can cause allergic reactions.	Practice sanitary milk collection procedures. Rapidly chill milk and hold at < 45°F (7°C). Maintain healthy herd, document milk withdrawal times, perform antibiotic residue testing.	1, 2
Raw milk storage	Biological: Pathogenic bacteria	Pathogenic bacteria and their toxins can cause illness.	Maintain milk temperature at < 40°F (4°C).	1
Bottling—jugs, caps, labels	Biological: Cross contamination from equipment Physical: Broken glass, metal shards	Pathogenic bacteria and their toxins can cause illness. Glass and metal shards pose a risk if ingested.	Bottling must be completed without milk temperature rising over 45°F (7°C). Equipment is thoroughly cleaned, sanitized, and inspected before and after use. Inspect bottles before using and after packaging.	1, 4
Storage after bottled	Biological: Growth of pathogenic bacteria, cross contamination with untested milk	Pathogenic bacteria and their toxins can cause illness.	Maintain storage temperature < 40°F (4°C). Store separately from untested and unbottled product.	1
Distribution	Biological: Growth of pathogenic bacteria	Pathogenic bacteria and their toxins can cause illness.	Do not sell milk that doesn't meet bacteriological standards; maintain temperature < 45°F (7°C) during transport and at market; ensure milk is properly labeled with required warning and that warning is also posted at farmers' market.	1, 3, 4

SAMPLE CHART OF CRITICAL CONTROL POINTS FOR RAW-MILK PRODUCTION

CCP #	Hazard	Critical Limits	Action	Records	Who
1 Milk temperature	Biological: Pathogenic bacteria	< 45°F (7°C)	Reject if over 45°F (7°C)	Batch record	Bottler
2 Antibiotic residue	Chemical: Allergic reaction to antibiotics	None	Reject if milk withholding times not followed	Batch record	Bottler/ milker
3 Bacteriological testing	Biological: Pathogenic bacteria	APC < 1500, total coliforms < 5, *E.coli* zero	Reject if limits not met	Batch records, milk bottling room log	Bottler
4 Metal/glass	Physical: Metal or glass shards	None	Reject milk if equipment or bottles show breakage	Batch record, milk bottling room log	Bottler
5 Warning label	Biolological: Pathogenic bacteria	Label read by customer	Ensure legible, easy-to-read label	Batch	Bottler

APPENDIX C

Sample Milk Purchase Agreement

This agreement is entered into on ______________________________[Date] between ________________________[Buyer] and ________________________[Seller].

1. This agreement shall be effective from __________________[Date] through _____________[Date]. The parties named above may, however, elect to renew this agreement for another term.
2. Buyer agrees to purchase ________[Species] milk not older than ____ days from Seller, to be used for the manufacture of cheese for the length of this agreement.
3. Buyer agrees to (a) pick up milk at the farm, or (b) Seller agrees to deliver milk to Buyer at a cost of $_____ per delivery.
4. Seller warrants that milk sold to Buyer shall be free of inhibitory substances and shall meet the standards set forth by the ______________ Department of Food and Agriculture for manufacturing of Grade _______ milk. Seller shall remain under state inspection throughout the life of the contract.
5. Milk must meet the approval of the Buyer based on any or all of the following tests:
   a) Taste and odor
   b) Bacteria*
      i. Coliforms not to exceed ______/ml (_____ to_____/ml payable at 90%; ______ to ______ payable at 80%)
      ii. SPC not to exceed ________/ml (_____ to_____/ml payable at 90%; ______ to ______ payable at 80%)
      iii. LPC not to exceed _______/ml (_____ to_____/ml payable at 90%; ______ to ______ payable at 80%)
      iv. PI not to exceed _____% of SPC (_____ to_____% payable at 90%; ______ to ______ payable at 80%)
      v. Other . . .
   c) Received at no less than 40°F/4°C (34°F/1°C preferred)
   d) Sediment (per dairy inspector's test)
   e) Any counts over the maximum high count agreed upon are subject to rejection by the Buyer. In the event of a high count that does not change product quality, the first two deductions taken will be 10 percent.
6. Weekly costs of testing milk and shipping to lab shall be divided equally between Buyer and Seller. Frequency: Milk will be tested _________ by an outside lab and sent in by Buyer. If milk-quality problems exist and more frequent testing is needed, Seller will pay costs and be responsible for shipping samples and proving to Buyer that milk meets quality standards before Buyer purchases milk again. Failure to produce milk that meets quality standards is grounds for cancellation of contract.

7. Buyer shall make no payment to Seller for milk that does not meet the conditions of paragraphs 4 and 5 above. Any payments made prior to testing the milk shall be credited to Buyer if milk does not meet same conditions.
8. If milk is shipped with inhibitory substances, or if milk is of such poor quality that cheese does not set up, the producer of that milk shall be financially liable for actual costs incurred by Buyer of labor, utilities, ingredients, transportation, and any other milk that was contaminated. Costs will be deducted from future milk checks. Seller will provide Buyer proof of liability insurance.
9. The price of milk will be based on a combination of butterfat, protein, and the time of year produced. The butterfat percentage × 0.66 + the protein percentage × 1.33 will yield a number which correlates to the payment schedule agreed upon. Buyer will make every effort to purchase all milk produced by Seller but cannot guarantee to purchase more than Seller shipped in the lowest quarter of the previous year. Buyer agrees to purchase all milk produced from Seller before adding new producers or purchasing from other outside sources. Seller agrees to give first option for purchase to Buyer for all marketable milk.
10. Buyer will pay a bonus of 10 percent for milk that has lab counts for point 5b, i–iii, at half or less. This means LPC and coliform counts less than ____ and raw counts less than ______.
11. Buyer agrees to renew this contract next year and purchase milk from Seller for another year provided Seller produces quality milk and similar amounts as in the previous year's lowest quarter.
12. Buyer and Seller agree that this agreement may be suspended in the event of acts of God or circumstances beyond the control of either party. Specifically, the agreement shall be suspended if Buyer dies or becomes disabled or incapacitated, either mentally or physically, so as to be unable to operate its business, or if the plant is partially or totally destroyed so as to cause a halt in the business. In addition, the agreement will be suspended if the laws of any governing body prohibit Buyer from manufacturing cheese.
13. Buyer and Seller agree that this agreement shall be nontransferable by either party without the written consent of both parties.

Attach agreed-upon milk purchase base price and payment schedule.

*Bacteria counts left blank, as some agreements are for Grade A or Grade B milk.

Contract adapted from one provided courtesy of Sarah Shevett, California.

APPENDIX D

Sample Floor Plans

Small home/hobby parlor and processing

Dimensions: 16′ × 16′.
Square Footage: 256.
Purpose: Production of dairy products for home use and private milk sales where legal.
Capacity: 10–20 does or ewes, 1–3 cows.
Notes: Plumbing on one interior wall for cost savings and freeze protection. Floor drains could be included for improved functionality and possible future licensing. Size can be easily altered (maintain 4'-interval dimensions for less building material waste and cost). Small on-demand hot water heater.
Designer: Gianaclis Caldwell.

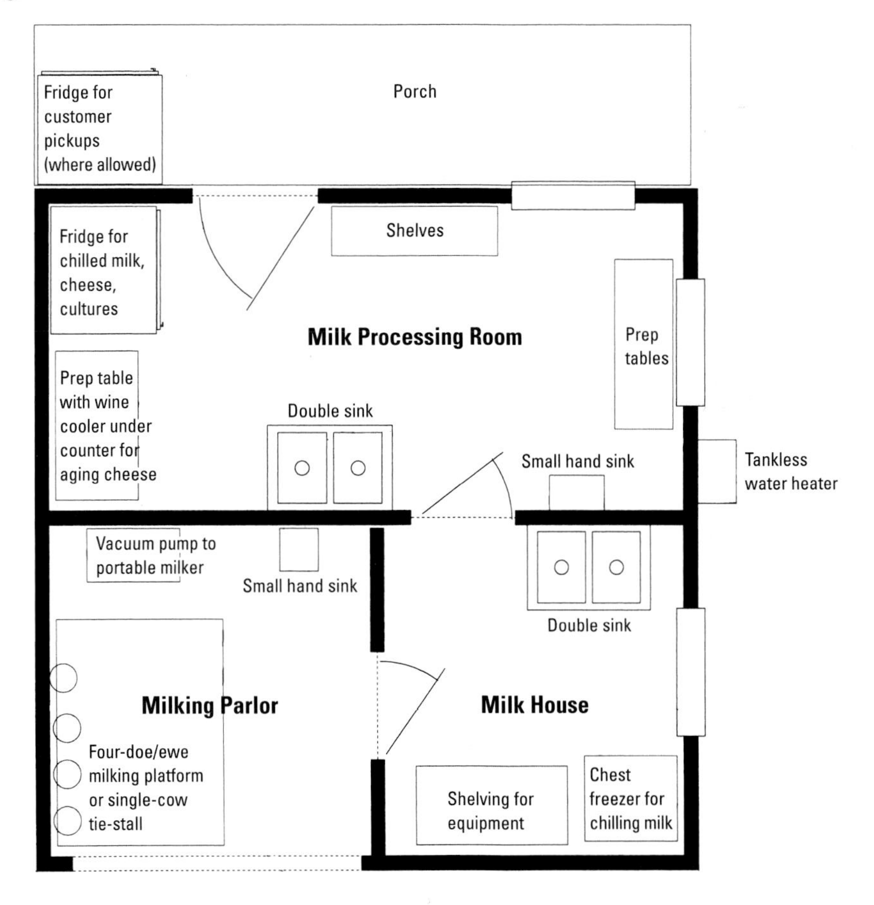

24′ × 36′ Dairy Barn

Dimensions: 24′ × 36′.
Square Footage: 864.
Purpose: Small commercial or herdshare dairy barn. (Confirm acceptance to regulatory requirements prior to construction.)
Capacity: 12–20 goats or 3–4 cows
Notes: Additional hay storage above or in separate barn. Separate maternity pen/barn.
Designer: Gianaclis Caldwell

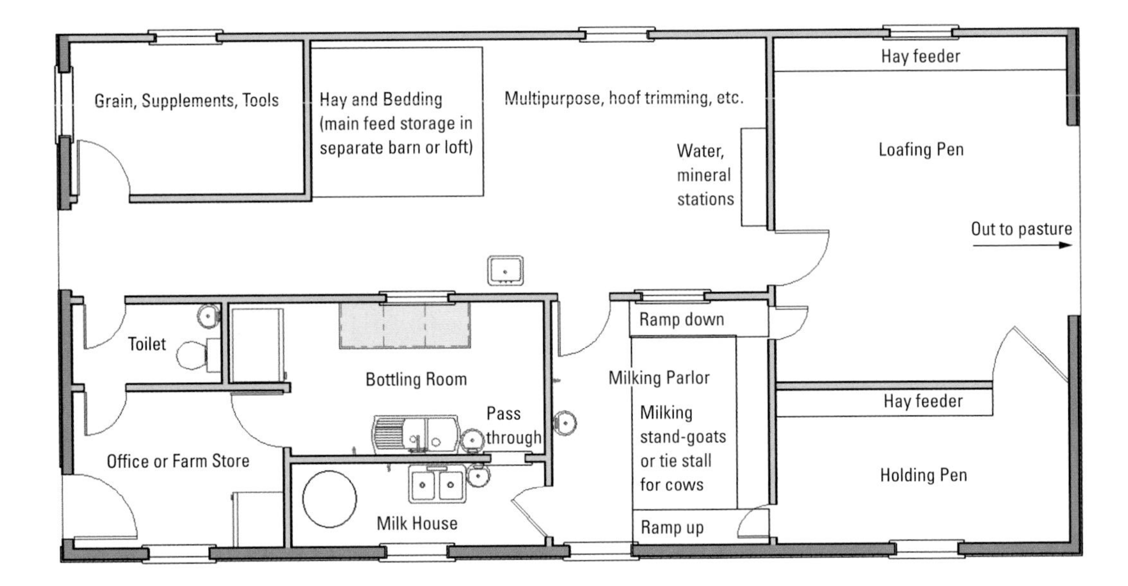

36′ × 60′ Dairy Barn

Dimensions: 36′ × 60′.
Square Footage: 2160.
Purpose: Medium-size commercial dairy barn (Confirm acceptance to regulatory requirements prior to construction.)
Capacity: 50–100 goats or 10–15 cows
Notes: Separate loafing, feeding, feed storage barns.
Designer: Gianaclis Caldwell

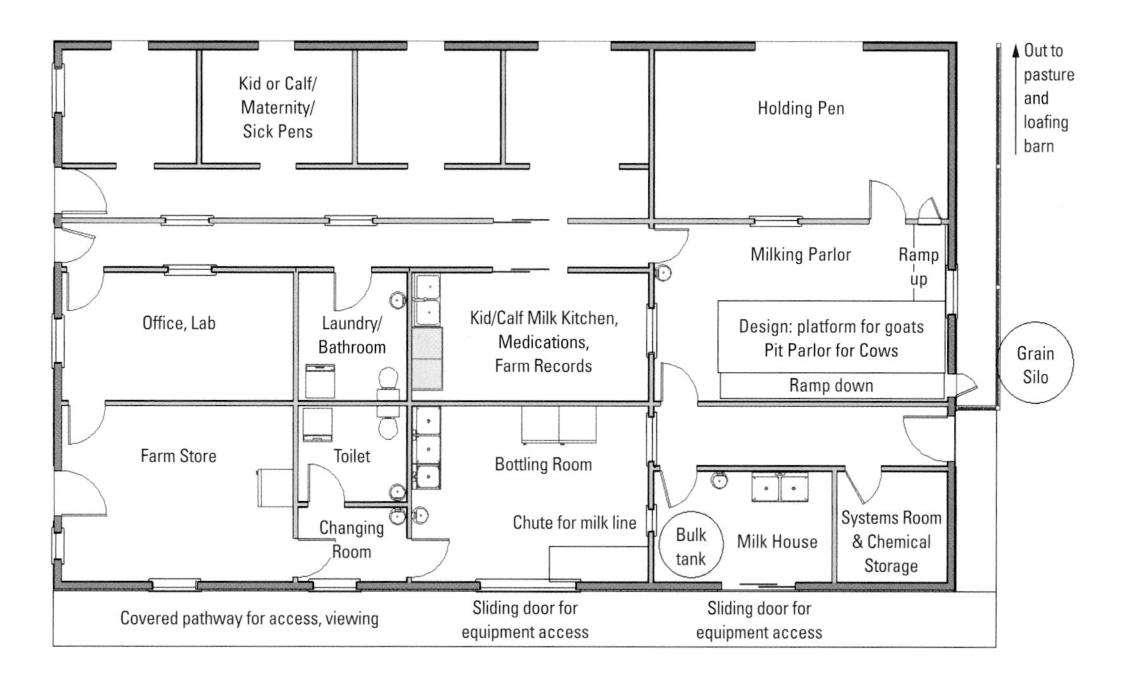

Glossary

Abomasum. The fourth compartment of a ruminant's upper digestive system, also known as the true stomach.

Aerobic. Oxygen is needed for metabolism.

Alpha casein. One of the main types of protein found in milk; includes several variants.

Anaerobic. No oxygen is needed for metabolism.

Apocrine. System in which a gland's secretions include portions of the secretory cells. Goats and humans secrete milk via the apocrine system (unlike cows and sheep).

A2 Milk. Milk that does not contain the genetic variant milk protein called A1 beta casein. Currently used to market milk produced by breeds and individual cows (that have been DNA typed) that do not produce the variant.

Batch or vat pasteurization. Lowest legal pasteurization temperature.

BCM-7. *See* **Beta casomorphin.**

Beta casein. The second most prevalent type of protein in milk; includes several variants.

Beta casomorphin. Also known as BCM. A protein fragment (peptide) that results from the breakdown of beta casein in milk. BCM is an opioid, a psychoactive chemical. BCM-7 contains seven amino acids.

Brewer's grain. Grain, usually barley, wheat, or oats, that are fermented as a part of the beer-making process. Spent brewer's grain refers to the wet low-sugar, high-protein grain that remains after draining.

Brucellosis. An infectious disease caused by the bacterium *Brucella.* Can be passed from animals to human through contact including raw milk. Formerly called undulant fever or Bang's disease.

Bulk tank. A receptacle that chills and holds milk, usually accumulated during more than one milking.

Buying clubs. Term used in this book to describe a collective that compensates another person to obtain and deliver their milk.

Casein. Main type of protein found in milk. Includes alpha, beta, and kappa caseins and their variants.

Certified dairy. Term that often refers to dairies that are certified (as differentiated from "licensed") by a regulatory body to produce and sell raw milk under the restrictions set forth by that body.

Clabber. To allow to sour, or to purposely sour, milk to the point that it coagulates.

Claw. Part of a milking machine that allows milk to pool before entering the hoses. Named after its resemblance to a bird claw (in the case of one designed to milk the four teats of a cow).

Coliforms. A large group of gas-forming bacteria that are found in soil, organic material, and the lower intestinal tracts of mammals.

Colostrum. The high-calorie, antibody-rich fluid secreted in the mammary gland of mammals at the time of giving birth and immediately thereafter.

Creaming. The behavior of fat globules in cow's milk that causes them to clump and rise to the top of the milk.

Decharacterize. To change the native character of a substance; in this book, refers to the changing of milk by the use of food coloring or charcoal to segregate its use for pets only.

Distillery grain. Grain remaining after the fermentation and distillation of alcohol, such as whiskey.

Enterohemorrhagic. Microbes that can cause bloody diarrhea and inflammation of the colon (colitis).

Enterovirulent. Microbes that can cause intestinal problems.

Enzymes. Proteins that accelerate biochemical processes such as the breakdown of a substance or the construction of another.

Facultative anaerobes. Microbes that prefer an oxygen-containing environment but that can adapt and survive without oxygen.

Fermentation. The metabolism of nutrients, such as sugars, by microbes with resulting by-products such as acid and alcohol.

Fertile Crescent. Area of the Middle East (also sometimes extended to the Nile River Delta and area) that is comparatively agriculturally fertile.

Fodder. Any feed for livestock, usually not including grains. Includes hay, chopped fresh grasses and legumes, etc.

Forage. Feed that is harvested by the animal, such as pasture for cows and sheep, and shrub-type matter for goats.

Forbs. Broad-leaved, nonwoody herbaceous plants, often called "weeds," that can provide nourishment to livestock.

Foremilk. The first few squirts of milk removed before milk collection. Foremilk contains higher numbers of bacteria that have made their way into the teat between milkings.

Galactose. One of two simple sugars that exist in milk as the disaccharide (double sugar) lactose, the other being glucose.

Glucose. One of two simple sugars that exist in milk as the disaccharide (double sugar) lactose, the other being galactose.

Grain. The seeds of grasses, including corn, fed to livestock as a source of nutrients and energy, primarily in the form of sugars (carboyhdrates).

Gram negative/positive. A technique for identifying certain categories of bacteria based on how their cell walls react to a certain type of stain.

Guillaine-Barre syndrome. Debilitating autoimmune disease often triggered by a previous illness such as an intestinal infection. Linked to campylobacter infection.

HACCP. Hazard Analysis Critical Control Point. A food risk reduction, or safety, program that focuses on identifying the points in a process at which a future risk/hazard can be reduced, limited, or eliminated.

Hand feeding. In this book, refers to feeding of infants by a means other than nursing at the breast. Term evolved before baby bottles and synthetic nipples were invented.

Hay. A dry fodder, usually of grasses or legumes, cut and stored for feeding livestock throughout the year. Often stored in tightly compacted bales.

Haylage. Hay that is purposely baled (compacted and secured) before completely drying and wrapped in plastic to allow lactic acid-producing bacteria to ferment and preserve.

Herd/cowshare or lease. Term used to describe a contractual agreement in which part or all of an animal or group of animals is owned by another individual. In this book the purpose being access to products from that animal or herd. Legally recognized in some states.

High-temperature short-time pasteurization (HTST). Heat treatment in which milk flows continuously through heating and cooling equipment, where it is heated to a minimum of 161°F (72°C) for 15 seconds, then rapidly cooled.

Homeostasis. A balanced condition; in this book it refers to a state of health that is in perfect balance.

Homogenization. A mechanical process that changes the behavior of (denatures) milk fat globules by reducing their size and altering the protein that encourages clumping of the globules, with the end result being that the milk fat will no longer rise to the surface.

Hypocalcemia. Also known as "milk fever," a condition that can occur shortly after a high producing cow, goat, or ewe gives birth, in which available calcium in the bloodstream drops to dangerously low levels, causing severe weakness and even death.

Indicator organism. Term describing a microorganism whose presence in a sample is indicative of the presence of other organisms.

Inflation. A soft rubber or silicone liner that is part of mechanical milking systems. The inflation makes direct contact with the teat.

Johnes. Pronounced "yo-knees." A chronic wasting disease found in ruminants; caused by *Mycobacterium avium* subspecies paratuberculosis. Also called paratuberculosis.

Lactase persistence. The genetic ability of a mammal past weaning age to produce the enzyme lactase, allowing them to digest the milk sugar lactose.

Lactic acid bacteria. Bacteria that metabolize, through fermentation, sugars to produce lactic acid bacteria.

Lactoperoxidase system. Natural system existing in varying degrees in fresh raw milk that enables milk to limit bacterial growth.

Lactose. Milk sugar consisting of the two simple sugars (monosaccharides) glucose and galatctose.

Lactose intolerance. Term describing varying degrees of the inability to digest the milk sugar lactose.

Lag phase. In this book, term applies to bacterial growth and the period of time before growth is observed.

Listeriosis. Disease state caused by pathogenic strains of the bacterium *Listeria.* Can be gastrointestinal or affect the central nervous system.

Mammary system. The milk-producing parts of a mammal; on ruminant milk animals includes the udder, teats, attachment tissue, and corresponding circulatory system.

Merocrine. System in which a gland secretes its substance without any loss of or damage to the gland itself. Cows and sheep secrete milk via the merocrine system (unlike humans and goats).

Mesophilic. Microbes that survive best at middle, warm temperatures.

Microbes/microorganisms. Term that applies to tiny microscopic organisms. Includes bacteria, fungi (yeasts and molds), viruses, and protozoa. Slang terms include "bugs" and "germs."

Milking zone. Refers to the area just above the milking pail or milking machine teat cups. Special attention should be given this area during milking hygienic steps.

Mortality/death phase. In this book, term applies to bacterial growth phase, occurring after stationary phase, when bacteria are beginning to expire.

Obligate aerobes. Microbes that need oxygen to survive.

Obligate anaerobes. Microbes that do not grow in oxygenated environments.

Omasum. The third chamber in a ruminant's upper digestive system.

Outbreak. With regard to food safety, refers to the occurrence of the identical infectious disease in more than one person within a similar time period.

Pasteurization. The process, named after Louis Pasteur, in which substances are exposed to heat to limit or destroy any present bacteria for the purposes of food safety or to extend shelf life.

Pathogen. A microbe that is capable of causing illness and disease.

Probiotic. Term that describes bacteria that provide a known benefit to health; namely, by their effect on the balance of microbes in the gut and through support of the immune system.

Psychrotrophes. Microbes that survive and multiply best in cooler temperatures.

Pulsator. Part of a mechanical milking system that provides pulsation of the vacuum system to the inflation (soft part that surrounds the teat during milking).

Q fever. Disease caused by the bacteria *Coxiella burnetii.* Animals can be infected and without disease symptoms. Passed to humans through multiple body fluids and by airborne means. Can be spread via raw milk.

Reticulum. The second chamber in a ruminant's upper digestive system.

Rickets. Nutritional deficiency caused by a lack of vitamin D, calcium, or phosphate.

Rumen. The first chamber of the upper digestive system of a ruminant (animal that chews a cud—cows, goats, sheep, deer, bison, etc.). The rumen ferments plant fiber.

Scurvy. Nutritional deficiency caused by a lack of vitamin C (ascorbic acid).

Shiga toxin. Powerful toxin produced by some bacteria. Named after the Japanese medical researcher who identified the dysentery-causing bacteria *Shigella dysenteriae* and the toxin it produces. Other bacteria, especially pathogenic strains of *E. coli*, also produce shiga toxin. Abbreviated Stx.

Silage. Chopped plant material such as corn, alfalfa, grain plant, and soybeans that is fermented as a feed for livestock.

Somatic cells, somatic cell count. Cells that originate from the animal's body and that are detectable in milk. Includes white blood cells whose presence over a certain number might indicate udder infection.

Stationary phase. Refers to phase of bacterial growth, after exponential growth phase, when growth levels out. Is followed by death/mortality phase.

Strip cup. Special cup that includes a strainer for use in observing the foremilk for abnormalities.

Stripping. The act of removing the foremilk or last milk, usually by hand.

Sub-therapeutic. The administration of a medication, often an antibiotic, at doses lower than needed to provide their intended therapy.

Swill milk. Term used to describe the substandard milk collected from cows fed waste grains from alcohol distillaries in the late nineteenth and early twentieth centuries.

Teat cup. Part of a mechanical milking system that includes a liner, called an inflation, and that is placed on the teat during milking.

Thermization. Heat treatments at temperatures lower than those required for pasteurization. Used for fluid milk and other products in which it is desired to reduce bacteria numbers with the least effect on other milk components.

Thermoduric. Microbes, often spore formers, that survive at higher temperatures (including some that survive pasteurization).

Thermophilic. Bacteria that survive and multiply at warm temperature. Commonly used in the making of yogurt.

Thrombotic *Thrombocytopenia purpura.* A life-threatening complication, including one in which small blood clots form throughout the body, blocking blood flow to vital organs such as the brain, kidneys, and heart.

Total mixed ration (TMR). A feed combined in a manner that is meant to supply the complete needs of the animal. Often includes silage, grains, minerals, and buffers.

Tuberculosis. Pulmonary disease caused by *Mycobacterium tuberculosis* but also can be transmitted to humans by *Mycobacterium bovis* through raw milk from an infected animal.

Ultra high temperature (UHT) pasteurization. Milk is heated to 280°F (138°C) for two seconds, then chilled rapidly.

Wet nurse. Term used to describe a woman hired or paid to provide breast milk to child not her own.

Bibliography

Chapter 1

Blayney, Donald, and Manchester, Alden C. "Milk Pricing in the United States." *Agriculture Information Bulletin.* No. (AIB-761). February 2001. Accessed from Economic Research Service, United States Department of Agriculture. Last updated May 26, 2012. http://www.ers.usda.gov/publications/aib-agricultural-information-bulletin/aib761.aspx#.Um6xDhbkr-A.

Carlson, Laurie Winn. *Cattle: An Informal Social History.* Chicago: Ivan R. Dee, 2001.

Dobbs, David. "Restless Genes." *National Geographic* (January 2013): 44.

Kindstedt, Paul. *Cheese and Culture.* White River Junction, Vt.: Chelsea Green Publishing, 2012.

Kozak, Jerry. "Rubbed Raw." National Milk Producers Federation. June 1, 2010. http://nmpf.org/latest-news/ceo-corner/jun-2010/rubbed-raw.

Mendelson, Anne. *Milk: The Surprising Story of Milk Through the Ages.* New York: Alfred A. Knopf, 2008.

Naaktgeboren, C. *The Mysterious Goat: Images and Impressions.* Eindhoven, The Netherlands: BB Press, 2006.

Valenze, Deborah. *Milk: A Local and Global History.* New Haven, Conn.: Yale University Press, 2011.

Woodford, Keith. *Devil in the Milk.* White River Junction, Vt.: Chelsea Green Publishing, 2007.

Chapter 2

Grant, Jennie P. *City Goats.* Seattle, Wash.: Mountaineers Books, 2012.

Gumpert, David E. *The Raw Milk Revolution: Behind America's Emerging Battle over Food Rights.* White River Junction, Vt.: Chelsea Green Publishing, 2009.

Real Raw Milk Facts. "Raw Milk Facts State By State." Accessed October 28, 2013. http://www.realrawmilkfacts.com/raw-milk-regulations.

Schmid, Ron. *The Untold Story of Milk: The History, Politics and Science of Nature's Perfect Food: Raw Milk from Pasture-Fed Cows.* Warsaw, Ind.: New Trends Publishing, 2009.

Chapter 3

A Campaign for Real Milk (project of the Weston A. Price Foundation). "The Facts About Real Raw Milk." Last updated February 28, 2013. http://www.realmilk.com.

Delaney, Carol. *A Guide to Starting a Commercial Goat Dairy.* Burlington: University of Vermont, 2012.

Chapter 4

Betsy, Tom, and Jim Keogh. *Microbiology Demystified.* New York: McGraw-Hill, 2012.

Emerging Infectious Diseases. "Nonpasteurized Dairy Products, Disease Outbreaks, and State Laws—United States, 1993–2006." http://www.cdc.gov/eid/, Vol. 18, No 3, March 2012.

Mullan, W. M. A. (2003) *Inhibitors in Milk.* Accessed January 10, 2013. http://www.dairyscience.info/inhibitors-in-milk/51-inhibitors-in-milk.html.

State of Washington Department of Agriculture. "Best Practices to Control Q Fever". Last updated August 5, 2011. http://agr.wa.gov/FoodAnimal/AnimalHealth/Diseases/QFeverManagementPractices.pdf.

Todar, Kenneth. "Bacterial Protein Toxins." *Todar's Online Textbook of Bacteriology* (website). Accessed October 28, 2013. http://textbookofbacteriology.net/proteintoxins.html.

World Health Organization. "Benefits and Potential Risks of the Lactoperoxidase System of Raw Milk Preservation." Report of an FAO/WHO technical meeting. Rome, Italy: November 28–December 2, 2005. http://www.who.int/foodsafety/publications/micro/lactoperoxidase/en/.

Food and Drug Administration. *Bad Bug Book: Foodborne Pathogenic Microorganisms and Natural Toxins.* Second Edition. 2012. http://www.fda.gov/Food/FoodborneIllnessContaminants/CausesOfIllnessBadBugBook/.

Chapter 5

Beals, Peg. *Safe Handling: Consumers' Guide, Preserving the Quality of Fresh, Unprocessed Whole Milk.* Chelsea, Mich.: Privately printed, 2011.

Dairy Practices Council. *Pre and Postmilking Teat Disinfectants.* EPC 49. Richboro, Pa.: Dairy Practices Council, 2000.

Goat Research. "E (Kika) de la Garza American Institute for Goat Research." Langston University, Langston, Okla. Last updated October 17, 2013. http://www.luresext.edu/goats/.

Karreman, Hubert J. *The Barn Guide to Treating Dairy Cows Naturally.* Austin, Tex.: Acres USA, 2011. This is my favorite dairy cow book for health issues and management. It is well illustrated and easy to read. Even if you are a goat breeder, you will learn something.

Laven, R. A., and K. E. Lawrence. "Efficacy of Blanket Treatment of Cows and Heifers with an Internal Teat Sealant in Reducing the Risk of Mastitis in Dairy Cattle Calving on Pasture." *New Zealand Veterinary Journal* 56 (4) (August 2008): 171–175. doi: 10.1080/00480169.2008.36830.

Lishman, A. W. "The Cow's Udder and Milk Secretion." Dairying in KwaZulu-Natal: KwaZulu-Natal Department of Agriculture and Environmental Affairs. Accessed October 28, 2013. http://agriculture.kzntl.gov.za/publications/production_guidelines/dairying_in_natal/dairy6_1.htm.

Philpot, W. Nelson, and Stephen C. Nickerson. *Winning the Fight Against Mastitis.* Naperville, Ill.: Westphalia Surge, 2000.

Chapters 6 and 7

Acres USA, http://www.acresusa.com/. Many books and publications, as well as a periodical, on holistic farm practices.

Bower, R. *Rumen Physiology and Rumination.* Last updated November 2009. Fort Collins, Colo.: Colorado State University. www.vivo.colostate.edu /hbooks/pathphys/digestion/herbivores/rumination.html.

Coleby, Pat. *Natural Goat Care.* Austin, Tex.: Acres USA, 2001.

Ensminger, M. E. *Dairy Cattle Science*, 2nd ed. Danville, Ill.: Interstate Printers and Publishers, 1980.

Ensminger, M. E. and Howard Tyler. *Dairy Cattle Science*, 4th ed. Upper Saddle River, N.J.: Prentice Hall, 2005.

Grandin, Temple. *Humane Livestock Handling: Understanding Livestock Behavior and Building Facilities for Healthier Animals.* North Adams, Ma.: Storey, 2008.

Hancock, Dale, and Tom Besser. *E. coli 0157:H7 in Hay or Grain-fed Cattle.* October 12, 2006. http://www.puyallup.wsu.edu/dairy/nutrient -management/publications.asp.

Lazor, Jack. *The Organic Grain Grower: Small-Scale, Holistic Grain Production for the Home and Market Producer.* White River Junction, Vt.: Chelsea Green Publishing, 2013.

Logsden, Gene. *Holy Shit: Managing Manure to Save Mankind.* White River Junction, Vt.: Chelsea Green Publishing, 2010. This book is both irreverent and purposeful—teaching and illuminating.

Morrow, Rosemary. *Earth User's Guide to Permaculture.* Hampshire, UK: Permanent Publications, 2010.

Turner, Newman, *Fertility Pastures.* Austin, Tex.: Acres U.S.A, 1955, 2009. A classic, ahead of its time on the relationship between soil fertility and animal health.

USDA Natural Resources Conservation Service. *Agricultural Waste Management Field Handbook.* NEH Part 651. Washington, D.C.: USDA Natural Resources Conservation Service. http://go.usa.gov/KoB.

Wightman, Tim. *Raw Milk Production Handbook.* Falls Church, Va.: Farm-to-Consumer Legal Defense Fund, 2008.

Chapters 8 and 9

Caldwell, Gianaclis. *The Small-Scale Cheese Business.* White River Junction, Vt.: Chelsea Green Publishing, 2010.

Dairy Practices Council. *Dairy Plant Sanitation.* DPC 57. Richboro, Pa.: Dairy Practices Council, 1998.

Dairy Practices Council. *Fundamentals of Cleaning and Sanitizing Farm Milk Handling Equipment.* DPC 9. Richboro, Pa.: Dairy Practices Council, 2005.

3A Sanitary Standards. Last updated December 4, 2013. http://www.3-a.org.

Chapter 10

Amagliani, G., A. Petruzzelli, E. Omiccioli, F. Tonucci, M. Magnani, and G. Brandi. "Microbiological Surveillance of a Bovine Raw Milk Farm through

Multiplex Real-Time PCR." *Foodborne Pathogens and Disease* 9(5) (2012): 406–11. doi: 10.1089/fpd.2011.1041.

Dairy Practices Council. *Raw Milk Quality Tests.* DPC 21. Richboro, Pa.: Dairy Practices Council, 2003.

Dairy Practices Council. *Troubleshooting High Bacteria Counts of Raw Milk.* DPC 24. Richboro, Pa.: Dairy Practices Council, 2001.

Lemire, Geneviéve. *Assessment of Milk Quality and Dairy Herd Health under Organic Management* (report). March 20, 2007. http://www.organicagcentre.ca/DOCs/Agri-reseau/Evaluation_Lait_Sante_en.pdf.

3M. *3M Petrifilm Interpretation Guide, Aerobic Count Plate.* St. Paul, Minn.: 3M, 2005. http://www.3m.com/intl/kr/microbiology/p_aerobic/use3.pdf.

3M. *3M Petrifilm Interpretation Guide, Coliform Count Plate.* St. Paul, Minn.: 3M, 1999. http://www.3m.com/intl/kr/microbiology/p_coliform/use3.pdf.

3M. *3M Petrifilm Interpretation Guide, E. coli/Coliform Count Plate.* St. Paul, Minn.: 3M, 2001. http://www.3m.com/intl/kr/microbiology/p_ecoli/use3.pdf.

Chapter 11

Dairy Practices Council. *Conducting and Documenting HACCP—Principle Number One: Hazard Analysis.* DPC 92. Richboro, Pa.: Dairy Practices Council, 2003.

Dairy Practices Council. *Conducting and Documenting HACCP—Principles #2 and 3, Critical Control Points and Critical Limits.* DPC 93. Richboro, Pa.: Dairy Practices Council, 2006.

Dairy Practices Council. *Conducting and Documenting HACCP—Principles #4 and 5, Monitoring and Corrective Action.* DPC 94. Richboro, Pa.: Dairy Practices Council, 2010.

Dairy Practices Council. *Conducting and Documenting HACCP, SSOP's and Prerequisite.* DPC 91. Richboro, Pa.: Dairy Practices Council, 2003.

Dairy Practices Council. *Good Manufacturing Practices for Dairy Processing Plants.* Dairy Practices Council. DPC 8. Richboro, Pa.: 1995.

Dixon, Peter. *HACCP-Based Program Handbook.* Westminster West, Vt.: Dairy Foods Consulting, 2012.

U.S. Food and Drug Administration. CDC Recall Plan templates. Last updated December 14, 2011. http://www.fda.gov/Safety/Recalls/IndustryGuidance/default.htm.

Chapter 12

Amrein-Boyes, Debra. *200 Easy Homemade Cheese Recipes.* Toronto: Robert Rose Inc., 2009.

Katz, Sandor Ellix. *The Art of Fermentation.* White River Junction, Vt.: Chelsea Green Publishing, 2012.

Mendelson, Anne. *Milk: The Surprising Story of Milk Through the Ages.* New York: Alfred A. Knopf, 2008.

Index

Note: Page numbers followed by f or t refer to Figures/Photographs or Tables

AMELIA CALDWELL

About the Author

Gianaclis Caldwell grew up on a small family farm in Oregon, where she milked cows, ran a dairy-cow 4-H club, and learned to raise organic produce and meat. In 2005, Gianaclis returned with her husband and their two daughters to the property, where they now operate Pholia Farm, an off-grid, raw-milk cheese dairy.

Caldwell is the author of *The Small-Scale Cheese Business*, a guide to building and running a small, on-farm creamery, and *Mastering Artisan Cheesemaking*, which won a 2013 Foreword Book of the Year Award in reference and was an International Association of Culinary Professionals Awards finalist. Her photographs and writing appear regularly in numerous publications, including *Culture* magazine.

Chelsea Green Publishing is committed to preserving ancient forests and natural resources. We elected to print this title on 10-percent recycled paper, processed chlorine-free. As a result, for this printing, we have saved:

8 Trees (40' tall and 6-8" diameter)
3,655 Gallons of Wastewater
4 million BTUs Total Energy
245 Pounds of Solid Waste
674 Pounds of Greenhouse Gases

Chelsea Green Publishing made this paper choice because we are a member of the Green Press Initiative, a nonprofit program dedicated to supporting authors, publishers, and suppliers in their efforts to reduce their use of fiber obtained from endangered forests. For more information, visit www.greenpressinitiative.org.

Environmental impact estimates were made using the Environmental Defense Paper Calculator. For more information visit: www.papercalculator.org.

the politics and practice of sustainable living

CHELSEA GREEN PUBLISHING

LIFE, LIBERTY, AND THE PURSUIT OF FOOD RIGHTS
The Escalating Battle Over Who
Decides What We Eat
DAVID E. GUMPERT
9781603584043
Paperback • $19.95

COWS SAVE THE PLANET
And Other Improbable Ways
of Restoring Soil to Heal the Earth
JUDITH D. SCHWARTZ
9781603584326
Paperback • $17.95

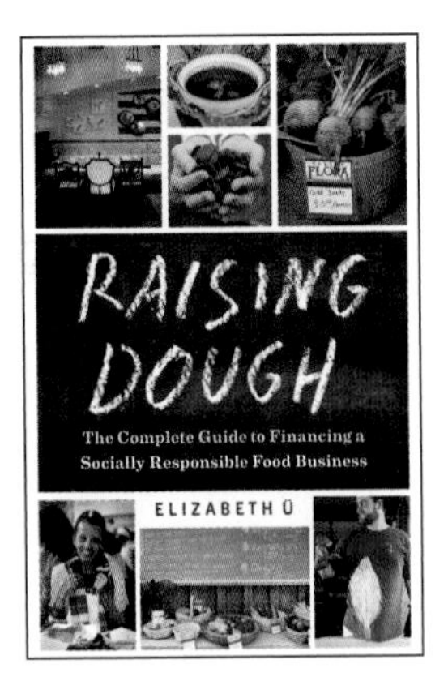

RAISING DOUGH
The Complete Guide to Financing
a Socially Responsible Food Business
ELIZABETH Ü
9781603584289
Paperback • $19.95

THE ART OF FERMENTATION
An In-Depth Exploration of Essential
Concepts and Processes from Around the World
SANDOR ELLIX KATZ
9781603582865
Hardcover • $39.95